Dynamical Systems Generated by Linear Maps

Ćemal B. Dolićanin · Anatolij B. Antonevich

Dynamical Systems Generated by Linear Maps

Second Edition

Ćemal B. Dolićanin
State University of Novi Pazar
Novi Pazar
Serbia

Anatolij B. Antonevich
Belarusian State University
Minsk
Belarus

ISBN 978-3-319-35830-7   ISBN 978-3-319-08228-8   (eBook)
ISBN 978-86-7466-489-6
DOI 10.1007/978-3-319-08228-8

Springer Cham Heidelberg New York Dordrecht London

ISBN 978-86-86893-47-5   1st edition: © State University of Novi Pazar 2012

Printed on acid-free paper

Springer is part of Springer Science+Business Media (www.springer.com)

Preface

The book deals with dynamical systems, generated by linear mappings of finite dimensional spaces. These systems have a relatively simple structure from the point of view of the modern dynamical systems theory. However, for dynamical systems of this kind, it is possible to obtain explicit answers to specific questions that are useful in applications.

The central place is occupied by the following problem. Let $A : \mathbb{C}^m \to \mathbb{C}^m$ be a linear nondegenerate mapping. The question is to investigate the trajectory of an arbitrary vector x under the action of the powers of linear mapping, that is, to give a description of the behavior of the sequence of vectors $A^n x$. In essence, this is a problem of investigation of the dynamical system given by linear mapping. These objects and questions arise in a series of applied and theoretical issues, first of all in iteration processes theory, and there are a significant number of publications on the theme.

At the same time linear mapping $A : \mathbb{C}^m \to \mathbb{C}^m$ generates a series of other dynamical systems. For example, under mapping A each d-dimensional subspace V is mapped to a d-dimensional subspace $A(V)$. The set of d-dimensional subspaces in $\mathbb{C}^m$ is the Grassmann manifold $G(m, d)$; thus the operator A generates the mapping φ_A of the Grassmann manifold $G(m, d)$ into itself, and generates a dynamical system with the phase space $G(m, d)$.

The monograph presents the results of investigation of dynamical systems of this kind and their applications.

The problems considered are natural and look rather simple (in essence their study is accessible to first-year university students), but in reality in the course of investigation, we are faced with a plenty of subtle questions and their detailed analysis needs a substantial effort.

The problems arising are related to linear algebra and dynamical systems theory, and therefore, the book can be considered as a natural amplification, refinement, and supplement to linear algebra and dynamical systems theory textbooks.

In Part I of the book the problem on the behavior of the sequence of vectors $A^n x$ is systematically considered for the first time. The main result is a complete decomposition of the trajectory of a vector, making it possible to obtain exhaustive information on a concrete trajectory (existence of the limit for $n \to +\infty$ and for $n \to -\infty$, description of the set of limit points and so on), as well as on the behavior of the direction vectors $\frac{1}{\|A^n x\|} A^n x$, and on the behavior of one-dimensional spaces generated by the vectors $A^n x$.

These results form a base for further investigation.

A linear mapping $A : \mathbb{C}^m \to \mathbb{C}^m$ generates a number of dynamical systems whose dynamics has not been previously investigated at all. The most important among these associated dynamical systems is the system generated by the afore-mentioned mapping φ_A on the Grassmann manifold $G(m, d)$. The question on the behavior of the trajectory $A^n(V)$ of a subspace V under the action of a linear mapping is also important in a number of problems, but previously there has been no investigation in this direction carried out. The complexity of the problem on the behavior of the trajectory of a subspace V is caused, first of all, by the fact that construction of the limit $\tilde{V}$ of such trajectory cannot be reduced to evaluation of the limits of trajectories of vectors or one-dimensional subspaces from V. The point is that in general, limits of trajectories of one-dimensional subspaces generate only a part of the limit subspace $\tilde{V}$.

In Part II, two approaches to the solution of this problem are proposed. These approaches are arbitrarily called algebraic and geometric. They are based on the original work of the authors and their collaborators.

The algebraic approach exploits the embedding of the Grassmann manifold into the projective space generated by the exterior power $\overset{d}{\bigwedge} X$ of the initial vector space. Under this embedding, the action of mapping φ_A is given by operator $\overset{d}{\bigwedge} A$—the exterior power of operator A. Thus the investigation of the trajectory of a d-dimensional subspace is reduced to investigation of the trajectory of a one-dimensional subspace under the action of a linear operator $\overset{d}{\bigwedge} A$. This allows us to apply the results of Part I.

The geometric approach rests on construction of linear combinations of vectors from subspace $A^n(V)$ that have nonzero limits. This enables one to construct the limit space. The mentioned linear combinations are built by means of the so-called renormalization on the base of a subtle analysis of the behavior of the trajectories of vectors.

The limit of the trajectory of a subspace is an invariant subspace. In the case of an operator with one point spectrum we have the most extensive family of invariant subspaces. Therefore, a description of the trajectory of a subspace under the action of an operator with one point spectrum, as well as evaluation of its limit, turned out to be the most difficult result.

Note that in the linear algebra and matrix theory there are plenty of subtle questions, and in spite of intensive investigation, a lot of answers to these questions are far from being answered, and the results obtained in this direction go far beyond the frames of the linear algebra courses and form the theme of special individual monographs. For example, the results on invariant subspaces for linear mappings of finite dimensional spaces comprise a voluminous monograph Gohberg I., Lancaster P., Rodman L. *Invariant subspace of matrices with applications*. New York–Chichester–Toronto–Singapore. Wiley & Sons. 1986.

The problems considered in the book belong to the family of these subtle questions. The results obtained can be applied in different areas of Mathematics. In Part III, one of the possible applications is presented. Here, we apply the results obtained for investigation of the spectral properties of weighted shift operators, being operators in functional spaces of the form

$$Bu(x) = a(x)u(\alpha(x)),$$

where α is a certain mapping of the domain of the functional space under consideration into itself.

In Chaps. 13 and 14, we consider weighted shift operators generated by linear mappings in the spaces of scalar functions. In their investigation, one naturally comes across a new class of dynamical systems, associated with linear mappings. These are dynamical systems having the phase space being a certain compactification of $\mathbb{R}^m$, where the action is given by an extension of the linear mapping onto this compactification.

On the basis of the results on the behavior of the trajectories of vectors, we obtain an explicit description of the spectrum and find the properties of operators $B - \lambda I$ for spectral values λ. In particular, the necessary and sufficient conditions for one-sided invertibility of such operators have been obtained. These conditions depend essentially on the dynamics of the corresponding extension of the linear mapping. Weighted shift operators in spaces of vector functions are more complexly structured and are more difficult in investigation. In Chap. 15, it is shown how in the investigation of weighted shift operators in spaces of vector functions the problems on behavior of the trajectories of subspaces arise. By means of model examples, it is demonstrated how the results on behavior of the trajectories of subspaces obtained in Part II help to derive an explicit description of spectral properties of weighted shift operators in spaces of vector functions.

In the course of investigation, we also discuss a series of auxiliary problems and structures. In particular, we present an original presentation of principal facts on exterior powers of vector spaces and the Grassmann manifolds.

In addition, we clarify relations between the problems under consideration and some classical mathematical theories such that the Lyapunov exponents theory, nonarchimedean normalization theory, vector bundles theory (in particular, the known Hopf bundle), commutative C^*-algebras and the Gelfand transform, graded-linear mappings, trajectories behavior on the torus under the standard shift action.

The authors have the pleasure to thank Professors Stevan Pilipovic, Peter Zabreiko, and Teodor Atanackovic, who carefully read the book and made a series of useful suggestions.

   The book was partially supported by the Serbian Ministry of Education, Science and Technological Development and Grant of the National Science Center (Poland) No DEC-2011/01/B/ST1/03838.

Ćemal B. Dolićanin
Anatolij B. Antonevich

Contents

Chapter 1
Introduction

1.1 Dynamical Systems

The general notion of a dynamical system encompasses the following components.

Phase space X. Points of this space are possible states of the system. One usually investigates phase spaces with some additional structures: X may be a measure space, a topological space or a smooth manifold.

Time. In the case of irreversible processes, the time may only be changed toward the future and it forms either the semigroup of non-negative real numbers (continuous time) $\mathbb{R}_0^+$, or the semigroup of non-negative integers $\mathbb{N}_0$. In the case of reversible processes, the time may be changed in both directions and it forms either the group $\mathbb{R}$ (continuous time), or the group $\mathbb{Z}$ of integers.

The law of system evolution. It is assumed that for each state $x \in X$ and each moment t of time, there is a state $F(x, t) \in X$, into which the system passes from the state x at the moment t. For each fixed t, one obtains a map $\varphi_t : x \mapsto F(x, t)$ of the phase space into itself. It is also assumed that the condition

$$\varphi_{t+s} = \varphi_t \circ \varphi_s,$$

which shows the stability of the process, has to be satisfied—the law of evolution does not depend on the origin of time. With x fixed, one obtains a map

$$t \mapsto F(x, t) \in X,$$

which is called a trajectory of the point x.

In this way, a reversible dynamical system with continuous time is a one-parameter group $\{\varphi_t \mid t \in \mathbb{R}\}$ of transformation of the space X into itself; a reversible dynamical system with discrete time is a group $\{\varphi_n \mid n \in \mathbb{Z}\}$ of bijective transformations of the space X into itself which has one generator $\varphi = \varphi_1$.

© Springer International Publishing Switzerland 2014
Ć.B. Dolićanin and A.B. Antonevich, *Dynamical Systems Generated by Linear Maps*,
DOI: 10.1007/978-3-319-08228-8_1

It is possible to consider a dynamical system with continuous time only at the moments of time of the form $t = nh$, where $h > 0$; one then obtains a new dynamical system with discrete time closely related to the original dynamical system.

It is usual to consider dynamical systems where the maps φ_t is consistent with the additional structure, in particular, it preserves the given measure, it is continuous or it is differentiable.

In the center of the attention in dynamical systems theory is a description of the asymptotic behavior of the system, i.e. the behavior of points and subsets of the phase space when times goes to infinity. Such a description uses existing structure on X. In particular, in the case of topological spaces, one of characteristics of a trajectory of a point x is the limit set

$$\Omega(x_0) = \{x \in X : \forall W(x) \; \forall N \in \mathbb{N} \; \exists n \geq N, \text{ such that } \varphi_n(x_0) \in W(x)\},$$

where $W(x)$ is neighborhood of the point x.

Among the questions considered in the theory of dynamical systems, let us emphasize the following:

1. How to determine those points x where the limit of the trajectory $\varphi_n(x)$ exists and how to find that limit in an explicit form?

2. What is the structure of the attracting set $U(x_0) = \{x| \; \varphi_n(x) \to x_0\}$?

3. If the limit of the trajectory $\varphi_n(x)$ does not exist, what is the structure of the set $\Omega(x)$ of the limit points of the trajectory and how to find $\Omega(x)$ explicitly?

1.2 Formulation of the Problems

Linear nonsingular transformation $A : \mathbb{C}^m \to \mathbb{C}^m$ generates a variety of dynamical systems.

1. First of all, the operator A itself generates a dynamical system acting on $X = \mathbb{C}^m$ by the powers A^n and one gets the problem of investigating the trajectory of an arbitrary vector x by the action of the linear transformation, i.e. the description of the behavior of the sequence of vectors of the form $A^n x$.

In addition to that, linear operator A generates dynamical systems on a string of other spaces related to $\mathbb{C}^m$.

2. For the description of the trajectory $A^n x$ of the vector x, it is natural to consider behavior of the "direction vectors" —the sequence of vectors

$$\xi_n = \frac{1}{\|A^n x\|} A^n x.$$

The vectors ξ_n lie on the sphere $S^{2m-1} = \{\xi \in \mathbb{C}^m| \; \|\xi\| = 1\}$ and the sequence ξ_n is the trajectory of the point ξ under the action of iterations of the mapping of the sphere into itself, given by the formula

$$\varphi : \xi \mapsto \frac{1}{\|A\xi\|} A\xi.$$

3. The set of one-dimensional subspaces of $\mathbb{C}^m$ is the complex projective space CP^{m-1}. Under a nonsingular transformation $A : \mathbb{C}^m \to \mathbb{C}^m$, a one-dimensional subspace passes into a one-dimensional subspace, therefore the operator A induces a map of CP^{m-1} into itself and generates a dynamical system with the phase space CP^{m-1}.

4. Similarly, the set of d-dimensional subspaces of $\mathbb{C}^m$ is the Grassmann manifold $G(m, d)$; under a nonsingular linear transformation a d-dimensional subspace V passes into a d-dimensional subspace $A(V)$, therefore the operator A induces a mapping φ_A of the manifold $G(m, d)$ into itself, which is called the Möbius transformation, and it generates a dynamical system with the phase space $G(m, d)$.

5. Let $X = \bigwedge^d \mathbb{C}^m$ be an exterior power of the space $\mathbb{C}^m$. An operator A induces a transformation $\bigwedge^d A$ of the space X and one gets a dynamical system with the phase space X.

6. There are various statements on dynamical systems holding in the case when X is a compact topological space. The space $\mathbb{C}^m$ is not compact and these results are not directly applicable. Even so, the space $\mathbb{C}^m$ may be embedded into different compact topological spaces X. Among various compactifications, there are spaces X such that the linear map extends to a continuous map of X into itself, and then the linear map induces a dynamical system with a compact phase space X.

Similar dynamical systems and similar questions arise in the real case, when the original operator A acts on the space $\mathbb{R}^m$. We do not discuss this case in detail, since the operator A extends to a linear operator of the space $\mathbb{C}^m$, and all results may be obtained from the results for the complex case.

In the aforementioned phase spaces, there are also dynamical systems with continuous time, generated by linear maps. The simplest one is related to the Cauchy problem for the system of ordinary differential equations with constant coefficients of the form

$$\frac{du}{dt} = Bu(t), \quad u(0) = x \in \mathbb{C}^m,$$

where B is a given matrix. As it is known, the solution is given by the formula

$$u(t) = e^{Bt} x,$$

and system of differential equations generates a dynamical system with continuous time, where the corresponding map φ_t is linear and it is given by the formula $\varphi_t(x) = e^{Bt} x$. In this example, the previously formulated questions are questions related to the asymptotic behavior of solutions of the system of differential equations.

In the study of solutions of the Cauchy problem for differential equations, it may occur that one only knows that the initial condition x in the Cauchy problem belongs to the given d-dimensional subspace V of $\mathbb{C}^m$, i.e. that the coordinates of the vector x in the initial moment of time satisfy some linear conditions. The question related to the subspace, containing values $u(t)$ of these solutions, for large t, leads to the discussion of dynamical systems on the Grassmann manifolds.

The need for research of the behavior of trajectories of subspaces, i.e. dynamical systems on the Grassmann manifolds, arises in a number of other problems from different areas, for example in the theory of linear extensions of dynamical systems, the theory of matrix factorization, the theory of normal forms of systems of differential equations, the theory of linear difference equations and weighted shift operators.

If there is a limit of the trajectory of the subspace V, then it is an invariant subspace for the operator A. So in studies of dynamical systems on the Grassmann manifolds, it is essential to know the form of invariant subspaces and the structure of the set of such subspaces. The consideration of the invariant subspaces of linear operators in finite-dimensional spaces raises a number of fairly complex issues, the results in this direction, known by 1985, are presented in the monograph [15] which is a volume of about 600 pages.

It is natural that the behavior of the trajectories depends on the properties operator A. For example, even in the simplest case of the operator acting on $\mathbb{R}^2$, there are many essentially different types of trajectories of vectors, the point 0 can be *node, focus, center, a saddle point*, etc. This example is ordinarily found in textbooks on differential equations and dynamical systems (see, for example, [1]).

The main results presented in the book, concerning the description of the dependence of the behavior of the trajectory on properties of the operator A, and the initial state—vector x or subspaces of V.

Dynamical systems generated by linear mappings have a relatively simple structure and many effects that are in the center of attention of the modern dynamical systems theory do not appear here. But, for these dynamical systems, it is possible to obtain explicit answers to specific questions which is useful in applications.

1.3 Some Classical Results

The question on the behavior of the sequence of vectors of the form $A^n x$ arises in many research areas, including the theory of iterative methods, the theory of Markov processes, and therefore, to numerous studies on specific problems. The summary of results in this direction is contained, for example, in [2].

One of the problems where the sequence of vectors of the form $A^n x$ arises is *modeling of evolution processes*.

In the simplest model of evolution processes, the state of the system is described by a vector x from a finite-dimensional subspace, and an evolution law is given by the matrix A: the state x in the unit of time passes into the state Ax. The problem is to describe the states of the system for large n, i.e. in the investigation of the behavior of the sequence of vectors $A^n x$.

In probability theory, one usually does not consider the whole vector space, but only the set of probability vectors and one considers stochastic matrices.

The vector $x = (x_1, \ldots, x_m)$ is called *probability vector*, if $x_i \geq 0$ and $\sum_{i=1}^{m} x_i = 1$.

The matrix A is called *stochastic*, if

$$a_{ij} \geq 0, \ \sum_{i=1}^{m} a_{ij} = 1.$$

Such probability process (which is usually called the *Markov chain* or *Markov process*) has the following interpretation.

One considers a system, containing many particles and each particle may be in one of m states. The component x_j of a probability vector gives the probability of finding the particle in the state with index j. The process consists of random transformations of particles from one state to another. The element a_{ij} of the stochastic matrix gives a probability of passing of the particle from the state i to the state j (in the unit of time). The condition of stochasticity is the expression of the condition that each particles passes into some state.

For a stochastic matrix, the vector $A^n x$ is a probability vector and it gives distribution of particles by states after n units of time. In particular, an eigenvector of the matrix A with eigenvalue 1 describes the *stationary state* of the system, which does not change over time.

One of the fundamental results in the theory of the Markov chains is the following theorem [2].

Theorem 1.3.1 (Markov) *If the matrix A is stochastic and $a_{ij} > 0$, then for eachs probability vector x, such that $x_i > 0$, the sequence $A^n x$ converges toward a probability vector y, the limit vector y does not depend on the initial vector x, and it is an eigenvector of the matrix A with eigenvalue 1.*

This proposition shows that for almost all initial conditions the system converges toward one (stationary) state. The condition $a_{ij} > 0$ reflects the fact of no forbidden transformations—the particle may from the state i pass to any other state.

In the Markov's theorem, the most important is the condition of positivity of elements of the matrix. For arbitrary positive matrices corresponding result was obtained by Perron.

Theorem 1.3.2 (Perron—Frobenius) *If all components of the matrix A are positive, then this matrix has a simple positive eigenvalue λ_0, which is larger than the modulus of any other eigenvalue. Moreover, for any vector x with positive coordinates the sequence of normalized vectors $\frac{1}{\|A^n x\|} A^n x$ converges to the eigenvector of the matrix A with eigenvalue λ_0.*

The condition of the positivity of matrix components is related to the basis in question. It is interesting to note, that the reversed proposition holds, i.e. the Perron theorem presents the complete characterization of operators which may be given by positive matrices.

Theorem 1.3.3 *Let the operator $A : \mathbb{R}^m \to \mathbb{R}^m$ has simple positive eigenvalue λ_0, which is larger than the modulus of any other eigenvalue. Then, there is a basis where all components of the matrix of that operator are positive.*

If the condition on positivity is not satisfied, the behavior of the trajectory becomes more complex and the sequence of normalized vectors $\frac{1}{\|A^n x\|} A^n x$ may not converge. One of the results which does not contain positivity condition was obtained for diagonal matrices.

Theorem 1.3.4 (Poincaré—Perron) *If the matrix A is diagonalizable and all eigenvalues have different moduli, then for every $x \neq 0$ there is such a numerical sequence γ_n, that the sequence $\gamma_n A^n x$ has the limit, and this limit is one of the eigenvectors of the matrix A.*

References

1. Katok, A., & Hasselblatt, B. (1998). *Introduction to the modern theory of dynamical systems*. Cambridge: Cambridge University Press.
2. Voevodin, V. V., & Kuznetsov, Y. A. (1984). *Matrix and calculations*. Moscow: Nauka.

Part I
Vector Trajectory

Chapter 2
The Jordan Basis and Special Subspaces

In this chapter Jordan basis and subspaces related to it will be discussed as well as various Jordan cells and basis vectors which play a different role at accidental numeration of separated basis vectors. The theory of subspace related with linear operators has been disposed.

2.1 The Special Numeration of the Jordan Basis

One calls the *Jordan cell* (of dimension q with eigenvalue λ) a matrix of the form

$$J = \begin{pmatrix} \lambda & 1 & \ldots & 0 & 0 \\ 0 & \lambda & \ldots & 0 & 0 \\ \ldots & \ldots & \ldots & \ldots & \ldots \\ 0 & 0 & \ldots & \lambda & 1 \\ 0 & 0 & \ldots & 0 & \lambda \end{pmatrix}. \tag{2.1.1}$$

Such a matrix may be represented in the form $J = \lambda I + N$ where I is the identity matrix, and matrix

$$N = \begin{pmatrix} 0 & 1 & \ldots & 0 & 0 \\ 0 & 0 & \ldots & 0 & 0 \\ \ldots & \ldots & \ldots & \ldots & \ldots \\ 0 & 0 & \ldots & 0 & 1 \\ 0 & 0 & \ldots & 0 & 0 \end{pmatrix}$$

is nilpotent: $N^q = 0$. Under the action of the matrix N we have $Ne_1 = 0$ and $Ne_k = e_{k-1}$ for $k > 1$. This means that the vector e_1 is an eigenvector with eigenvalue 0, and all other vectors e_k are adjoint. Note that in the representation of a linear operator in the form of the Jordan cell, numeration of the vectors from the basis is essential. The Jordan cell may be of dimension 1, with no adjoint vectors.

© Springer International Publishing Switzerland 2014
Ć.B. Dolićanin and A.B. Antonevich, *Dynamical Systems Generated by Linear Maps*,
DOI: 10.1007/978-3-319-08228-8_2

One also considers lower-triangular Jordan cells having the form

$$
J = \begin{pmatrix}
\lambda & 0 & \ldots & 0 & 0 \\
1 & \lambda & \ldots & 0 & 0 \\
\ldots & \ldots & \ldots & \ldots & \ldots \\
0 & 0 & \ldots & \lambda & 0 \\
0 & 0 & \ldots & 1 & \lambda
\end{pmatrix}.
$$

That matrix is similar to (2.1.1) and their difference is only in the method of numeration of the elements of the basis.

As we know, from the standard courses in linear algebra, if X is finite-dimensional complex vector space, then for any linear operator $A : X \rightarrow X$ there exists the *Jordan form*, i.e. there exists the basis (*Jordan basis*), with respect to which the matrix of the operator has block-diagonal form, with the Jordan cells on the diagonal. The Jordan form is uniquely determined up to the order of the Jordan cells, and these cells are usually numbered randomly. But, in a number of problems, including those discussed below, different Jordan cells and different basis vectors have different roles. In the case of random numeration, the extraction of basis vectors, needed for some concrete situation, is carried out usually with cumbersome verbal descriptions. The corresponding objects may be given by formulas, if one employs the following specific numeration of the vectors of such a basis, using four indices. This numeration of the vectors of the Jordan basis was introduced in [1].

The given a linear operator $A : X \rightarrow X$, where X is a m-dimensional complex vector space, denote by q the number of different moduli of eigenvalues of this operator, and numerate these moduli in the increasing order

$$0 \leq r_1 < r_2 < \cdots < r_q.$$

Let $q(k)$ be the number of eigenvalues with modulus r_k. Let us numerate all different eigenvalues with the given modulus r_k, using the second index j. We obtain a numeration of the set of different eigenvalues using the two indices k and j.

$$\{\lambda_{kj} : 1 \leqslant k \leqslant q, \ 1 \leqslant j \leqslant q(k)\}.$$

Here, for definiteness, we may assume that for any k, numbers λ_{kj} are numerated according to its location on the circle $|\lambda| = r_k$, i.e. in the increasing order with respect to the argument of the complex number. Otherwise speaking, the eigenvalues are presented in the form $\lambda_{kj} = r_k e^{ih_{kj}}$ where $0 \leq h_{kj} < 2\pi$, and we assume that the numeration is chosen in such a way that $h_{k,j} < h_{k,j+1}$. Note here that the method of numeration is not essential.

Let $q(k, j)$ be the number of Jordan cells corresponding to the eigenvalue λ_{kj}; for any admissible couple (k, j) numerate those cells by the third index i, and denote these cells by $J(k, j, i)$. We obtain a numeration of all Jordan cells by three indices:

$$\{J(k, j, i) : 1 \leqslant k \leqslant q; \ 1 \leqslant j \leqslant q(k), \ 1 \leqslant i \leqslant q(k, j)\}.$$

We also assume that for given k and j, the Jordan cells are numerated in the nondecreasing order with respect to their dimensions. The cells of equal dimensions with equal eigenvalues are of equal importance and they are numerated arbitrarily.

Let $q(k, j, i)$ be the dimension of the cell $J(k, j, i)$. We numerate all basis vectors corresponding to this cell, using the additional fourth index l, where here the numeration is uniquely determined by the structure of the Jordan cell, according to the rule formulated earlier.

In this way, vectors from the basis in which the matrix of the operator has the Jordan form, are naturally numerated using four indices:

$$e(k, j, i, l) : 1 \leqslant k \leqslant q; \ 1 \leqslant j \leqslant q(k), \ 1 \leqslant i \leqslant q(k, j), \ 1 \leqslant l \leqslant q(k, j, i).$$

This method of numeration is convenient since different indices represent different characteristics of the corresponding basis vector, and this numeration simplifies recording of constructions discussed below.

The coordinates of a vector x in the constructed bases will be denoted by $x(k, j, i, l)$.

We will write below the corresponding indices of basis vectors (and also of coordinates, operators, eigenvalues etc.) in parenthesis and not in the form of lower of upper indices. This notation is more convenient in the case where one uses indices of vectors given by complicated expressions in the considered formulas.

Proposition 2.1.1 *Under the chosen numeration of elements of the Jordan bases, the action of the operator A on the basis vectors is given by formulas:*

$$Ae(k, j, i, 1) = \lambda(k, j, i)e(k, j, i, 1);$$

$$Ae(k, j, i, l) = \lambda(k, j, i)e(k, j, i, l) + e(k, j, i, l - 1) \text{ if } l > 1.$$

In particular, in this notation, the vector $e(k, j, i, 1)$ is an eigenvector with the eigenvalue $\lambda(k, j)$, vectors $e(k, j, i, l)$ for $l > 1$ are adjoint vectors of order $l - 1$.

Without loss of generality, following this, it is assumed that there is fixed a scalar product on the given vector space such that the chosen basis vectors are orthonormal.

Let us note that for a given operator there are many bases for which the matrix of that operator has the Jordan form, and the numeration discussed above is related to the specific basis. Therefore, some objects (subspaces, projectors etc.) appearing in the following constructions, depend on the choice of the basis and some of them are canonical—they are uniquely determined by the operator itself and they do not depend on the choice of a basis. Naturally, the final results should not depend on the choice of a basis.

Let us give another remark on the notion of a matrix. Ordinarily, in the courses on linear algebra, by a matrix of dimension $m \times n$ one calls a collection of numbers $\{a_{ij}\}$, indexed by two indices $1 \leq i \leq m$, $1 \leq j \leq n$. In other words, a matrix is defined as a function on the product $M \times N$ of two sets $M = \{1, 2, \ldots, m\}$, $N = \{1, 2, \ldots, n\}$. A numeration of the elements of the sets M and N by natural numbers is convenient for the presentation of the matrix in the form of a table, on a piece of paper. But, in most cases, the actual numeration of elements of the sets M and N is of no importance and Bourbaki [2] defines a matrix as a function on the product $M \times N$ of two arbitrary finite sets. In particular, in the previous construction the set M, containing m elements, it is realized as the set consisting of 4-tuples of integers:

$$M = \{(k, j, i, l) : 1 \leqslant k \leqslant q; \ 1 \leqslant j \leqslant q(k), \ 1 \leqslant i \leqslant q(k, j), \ 1 \leqslant l \leqslant q(k, j, i)\}.$$

2.2 Subspaces Related to a Linear Operator

In the following, we need a number of subspaces generated by specific collections of basis vectors from the Jordan basis and projectors onto these spaces.

Let us denote by $P(k, j, i, l)$ the operator, acting by the formula

$$P(k, j, i, l)x = x(k, j, i, l)e(k, j, i, l).$$

This operator is the orthogonal projector onto the one-dimensional subspace $L(k, j, i, l)$, generated by the basis vector $e(k, j, i, l)$.

Let $L(k, j, i)$ be the subspace generated by vectors $e(k, j, i, l)$, with fixed k, j, i and arbitrary l. This subspace corresponds to the Jordan cell $J(k, j, i)$. By $P(k, j, i)$, we denote the orthogonal projector onto the subspace $L(k, j, i)$. We shall denote by $x(k, j, i)$ the projection of the vector x onto the subspace $L(k, j, i)$:

$$x(k, j, i) = P(k, j, i) \, x = \sum_l x(k, j, i, l)e(k, j, i, l).$$

Note that the subspaces $L(k, j, i)$ and projectors $P(k, j, i)$ we have just introduced are not canonical, since they depend on the choice of a basis, and in the general case, they are not uniquely determined for the given operator.

By $L(k, j)$, we denote the subspace generated by all the vectors $e(k, j, i, l)$, for fixed k and j. This subspace is independent of a choice of a basis and it is canonical—this is the so-called *root subspace* corresponding to the eigenvalue $\lambda(k, j)$ and it can be given directly using the operator A by:

$$L(k, j) = \ker(A - \lambda(k, j)I)^{p(k, j)},$$

where $p(k, j) = \max_i q(k, j, i)$.

Let us denote the projection of the vector x onto the subspace $L(k, j)$ by $x(k, j)$, and corresponding projector by $P(k, j)$:

$$x(k, j) = P(k, j) \, x = \sum_i \sum_l x(k, j, i, l)e(k, j, i, l).$$

Let
$$L(k) = \bigoplus_j L(k, j),$$

i.e. this is the subspace generated by all basis vectors $e(k, j, i, l)$, for a given k. These subspaces are also canonical—they do not depend on choice of a basis.

The operator
$$P(k) = \sum_j P(k, j)$$

is the orthogonal projector onto the subspace $L(k)$. The notion of the orthogonal projector is related to the given scalar product. Nevertheless, the operator $P(k)$ may be introduced as a projector onto the subspace $L(k)$, whose kernel contains all subspaces $L(j)$ for $j \neq k$, therefore the operators $P(k)$ are also canonical.

We denote by $x(k)$ the projection of a vector x onto the subspace $L(k)$:

$$x(k) = P(k)x = \sum_j \sum_i \sum_l x(k, j, i, l)e(k, j, i, l).$$

It is obvious that all subspaces $L(k)$, $L(k, j)$ and $L(k, j, i)$ are invariant with respect to the action of A, and there is a decomposition

$$X = \bigoplus_{k=1}^{q} L(k) = \bigoplus_{k=1}^{q}\bigoplus_{j=1}^{q(k)} L(k, j) = \bigoplus_{k=1}^{q}\bigoplus_{j=1}^{q(k)}\bigoplus_{i=1}^{q(k,j)} L(k, j, i). \qquad (2.2.2)$$

If by $A(k)$, $A(k, j)$, $A(k, j, i)$, one denotes the restriction of the operator A on the corresponding subspace, then

$$A = \bigoplus_{k=1}^{q} A(k) = \bigoplus_{k=1}^{q}\bigoplus_{j=1}^{q(k)} A(k, j) = \bigoplus_{k=1}^{q}\bigoplus_{j=1}^{q(k)}\bigoplus_{i=1}^{q(k,j)} A(k, j, i).$$

Let us also introduce a collection of subspaces

$$M(k) = \bigoplus_{s=1}^{k} L(s). \qquad (2.2.3)$$

These subspaces are also invariant with respect to the action of the operator A. In addition to it, it is obvious that

$$M(1) \subset M(2) \subset \cdots \subset M(q),$$

i.e. the subspaces $M(k)$ form a pyramid.

Note that the sequence $M = (0 = M_0, M_1, M_2, \ldots, M_p)$ of vector subspaces of the space X is called a *pyramid*, if $M_j \subset M_{j+1}$ and $M_p = X$ (one also uses the term *flag*) [3]. The subspaces M_j are called *elements of the pyramid*.

The set-theoretical difference $S_j = M_j \setminus M_{j-1}$ is called a *level* of the pyramid M, and number $\dim M_j - \dim M_{j-1}$ is called the *weight* of the level. The subspace $M_0 = \{0\}$ is called the null level with weight 0. The set of levels is ordered by the order relation, generated by the index of the level: a level with index j follows a level with index k if $j > k$.

In this terminology, each subspace M_j is the set-theoretical union of levels contained in it, the levels are disjoint and the dimension of the subspace M_j is the sum of weights of the levels contained in it.

For the given pyramid (2.2.3) and each vector x, let us introduce a number characterizing the location of the vector with respect to the pyramid: this number $k(x)$ is the index of the level $M(k) \setminus M(k-1)$, containing the given vector x. This number may also be described as the largest of the indices k of basis vectors $e(k, j, i, l)$, actually (with a non-zero coefficient) appearing in the decomposition of the vector x.

We also need some other subspaces.

For given k, j and l, let us introduce the subspace $L^l(k, j)$, generated by basis vectors $e(k, j, i, l)$, for given k, j, l and different i (if they exist). The projector onto the subspace $L^l(k, j)$ is denoted by $Q(k, j; l)$:

$$Q(k, j; l)x = \sum_i x(k, j, i, l)\mathbf{e}(k, j, i, l). \tag{2.2.4}$$

The subspace $L^1(k, j)$ is generated by eigenvectors with eigenvalues $\lambda(k, j)$, the subspace $L^2(k, j)$ is generated by adjoint vectors of the first order, etc.

Dimensions of subspaces $L^l(k, j)$ do not increase with the increase of l:

$$\dim L^{l+1}(k, j) \leq \dim L^l(k, j)),$$

and the number of such subspaces is the number $p(k, j) = \max_i \{q(k, j, i)\}$—the largest dimension of the Jordan cells with eigenvalue $\lambda(k, j)$.

Let us also introduce the subspace

$$L^l(k) = \bigoplus_j L^l(k, j)$$

and corresponding projector $Q(k; l)$:

$$Q(k; l)x = \sum_j Q(k, j; l)x = \sum_j \sum_i x(k, j, i, l)\mathbf{e}(k, j, i, l). \tag{2.2.5}$$

Obviously, one has decompositions

$$L(k, j) = \bigoplus_{l=1}^{p(k,j)} L^l(k, j),$$

$$L(k) = \bigoplus_{l=1}^{p(k)} L^l(k), \quad p(k) = \max_j p(k, j).$$

In the following, the fundamental role will be played the family of subspaces indexed by two indices (k, s), where $s \le p(k)$:

$$M^s(k) = \left[\bigoplus_{i=1}^{k-1} L(i)\right] \bigoplus \left[\bigoplus_{l=1}^{s} L^l(k)\right] = M(k-1) \bigoplus \left[\bigoplus_{l=1}^{s} L^l(k)\right]. \quad (2.2.6)$$

These subspaces are ordered by inclusion

$$M^s(k) \subset M^{s_1}(k_1) \iff \text{if } k < k_1 \text{ and if } k = k_1, s \le s_1.$$

This ordering induces *lexicographic ordering* on the set of pairs (k, s):

$$(k_1, s_1) \preccurlyeq (k_2, s_2), \text{ if } k_1 < k_2 \text{ and if } k_1 = k_2 \text{ and } s_1 \le s_2. \quad (2.2.7)$$

In this way, the pyramid formed by these subspaces has the form

$$M^1(1) \subset M^2(1) \subset \cdots M^{p(1)}(1) \subset M^1(2) \subset M^2(2) \subset \cdots \subset M^{p(q)}(q). \quad (2.2.8)$$

Note that $M^{p(k)} = M(k)$, i.e. the pyramid (2.2.3) is a part of the pyramid (2.2.8). The pyramid formed by subspaces $M^s(k)$ shall called the *thickened pyramid* generated by operator A.

The location of the vector x, with respect to the pyramid, will be characterized by the pair of numbers $(k(x), s(x))$—the index of the level of the thickened pyramid (2.2.8), containing the vector x. As it will be shown below, this pair of numbers plays a fundamental role in the description of the behavior of the trajectory of the vector x.

References

1. Antonevich, A., & Buraczewski, A. (1996). Dynamics of linear mapping and invariant measures on sphere. *Demonstratio Mathematica*, 29(4), 817–824.
2. Bourbaki, N. (1962). *Éléments de Mathématiques, Algebre*. Paris: Hermann.
3. Bylov B. F., Vinograd R. E., Grobman D. M., & Nemyckii V. V. (1966). Lyapunov exponents. Moscow: Nauka.

Chapter 3
Representation of the Vector Trajectory

As already noted, in various applications, there are different questions on the behavior of a sequence of vectors of the form $A^n x$. In some cases, in order to answer such questions, it is enough to construct the leading term of the asymptotic of the trajectory $A^n x$, and this has been done in many papers (see for example [1]). However, in more subtle investigations, in particular, for the description of the dynamic of a linear map on the Grassmann manifold, all terms of the asymptotic expansion of the trajectory of an arbitrary vector are necessary.

In this chapter, an explicit representation of the trajectory $A^n x$ of an arbitrary vector is obtained, and the description of the behavior of such trajectory is given. The formulation of final results uses many auxiliary characteristics and operators naturally occuring in the process of investigation.

In the following, the basis for which the operator A has the Jordan form (with the numeration by four indices from Sect. 2.1), is given and the previously introduced special subspaces and the notation is used either.

3.1 Functions of the Jordan Cell

As it is known [2], the degree of the Jordan cell of the form (2.1.1) and dimension m is given by the formula

$$
J^n = \begin{pmatrix}
\lambda^n & C_n^1 \lambda^{n-1} & C_n^2 \lambda^{n-2} & \cdots & C_n^{m-2} \lambda^{n-m+2} & C_n^{m-1} \lambda^{n-m+1} \\
0 & \lambda^n & C_n^1 \lambda^{n-1} & \cdots & C_n^{m-3} \lambda^{n-m+3} & C_n^{m-2} \lambda^{n-m+2} \\
0 & 0 & \lambda^n & \cdots & C_n^{m-4} \lambda^{n-m+4} & C_n^{m-3} \lambda^{n-m+3} \\
0 & 0 & 0 & \cdots & C_n^{m-5} \lambda^{n-m+5} & C_n^{m-4} \lambda^{n-m+4} \\
\vdots & \vdots & \vdots & \ddots & \vdots & \vdots \\
0 & 0 & 0 & \cdots & \lambda^n & C_n^1 \lambda^{n-1} \\
0 & 0 & 0 & \cdots & 0 & \lambda^n
\end{pmatrix}, \qquad (3.1.1)
$$

© Springer International Publishing Switzerland 2014
Ć.B. Dolićanin and A.B. Antonevich, *Dynamical Systems Generated by Linear Maps*,
DOI: 10.1007/978-3-319-08228-8_3

where

$$C_n^k = \frac{n!}{k!(n-k)!}$$

are binomial coefficients.

The formula (3.1.1) may easily be obtained directly by straightforward calculations. Namely, we have $J = \lambda I + N$, where the summands commute. Therefore, by the binomial formula

$$J^n = (\lambda I + N)^n = \sum_{s=0}^{n} C_n^s N^s \lambda^{n-s}.$$

Since $N^s = 0$ for $s \geq m$, in the last sum only m summands remain:

$$J^n = \sum_{s=0}^{m-1} C_n^s N^s \lambda^{n-s}.$$

The equality (3.1.1) is the record of the last equation in the matrix form.

Note that if $\lambda = 0$, the formula simplifies: we have $J = N$ and $J^n = 0$ for $n \geq m$.

The equality (3.1.1) is a partial case of the following general proposition [2].

Proposition 3.1.1 *If the function f is defined on the neighborhood of the point λ on the complex plane and it is m times differentiable, then for the Jordan cell J the matrix $f(J)$ can be defined by the formula*

$$f(J) = \begin{pmatrix} f(\lambda) & \frac{1}{1!}f^{(1)}(\lambda) & \frac{1}{2!}f^{(2)}(\lambda) & \cdots & \frac{1}{(m-1)!}f^{(m-1)}(\lambda) \\ 0 & f(\lambda) & \frac{1}{1!}f^{(1)}(\lambda) & \cdots & \frac{1}{(m-2)!}f^{(m-2)}(\lambda) \\ 0 & 0 & f(\lambda) & \cdots & \frac{1}{(m-3)!}f^{(m-3)}(\lambda) \\ 0 & 0 & 0 & \cdots & \frac{1}{(m-4)!}f^{(m-4)}(\lambda) \\ \vdots & \vdots & \vdots & \ddots & \vdots \\ 0 & 0 & 0 & \cdots & \frac{1}{1!}f^{(1)}(\lambda) \\ 0 & 0 & 0 & \cdots & f(\lambda) \end{pmatrix}. \tag{3.1.2}$$

For example, in the case $f(z) = e^{tz}$ we have

$$e^{Jt} = \begin{pmatrix} e^{\lambda t} & \frac{1}{1!}te^{\lambda t} & \frac{1}{2!}t^2 e^{\lambda t} & \cdots & \frac{1}{(m-2)!}t^{m-2}e^{\lambda t} & \frac{1}{(m-1)!}t^{m-1}e^{\lambda t} \\ 0 & e^{\lambda t} & \frac{1}{1!}te^{\lambda t} & \cdots & \frac{1}{(m-3)!}t^{m-3}e^{\lambda t} & \frac{1}{(m-2)!}t^{m-2}e^{\lambda t} \\ 0 & 0 & e^{\lambda t} & \cdots & \frac{1}{(m-4)!}t^{m-4}e^{\lambda t} & \frac{1}{(m-3)!}t^{m-3}e^{\lambda t} \\ 0 & 0 & 0 & \cdots & \frac{1}{(m-5)!}t^{m-5}e^{\lambda t} & \frac{1}{(m-4)!}t^{m-4}e^{\lambda t} \\ \vdots & \vdots & \vdots & \ddots & \vdots & \vdots \\ 0 & 0 & 0 & \cdots & e^{\lambda t} & \frac{1}{1!}te^{\lambda t} \\ 0 & 0 & 0 & \cdots & 0 & e^{\lambda t} \end{pmatrix}.$$

Since $A^{-1} = f(A)$, where $f(z) = \frac{1}{z}$, and $f^{(k)}(\lambda) = (-1)^k k! \frac{1}{\lambda^{k+1}}$, for the inverse matrix (under condition $\lambda \neq 0$) we have the expression

$$
J^{-1} = \begin{pmatrix}
\frac{1}{\lambda} & -\frac{1}{\lambda^2} & \frac{1}{\lambda^3} & \cdots & (-1)^m \frac{1}{\lambda^m} & (-1)^{m+1} \frac{1}{\lambda^{m+1}} \\
0 & \frac{1}{\lambda} & -\frac{1}{\lambda^2} & \cdots & \frac{1}{(m-3)!} t^{m-3} e^{\lambda t} & (-1)^m \frac{1}{\lambda^m} \\
0 & 0 & \frac{1}{\lambda} & \cdots & \frac{1}{(m-4)!} t^{m-4} e^{\lambda t} & \frac{1}{(m-3)!} t^{m-3} e^{\lambda t} \\
0 & 0 & 0 & \cdots & \frac{1}{(m-5)!} t^{m-5} e^{\lambda t} & \frac{1}{(m-4)!} t^{m-4} e^{\lambda t} \\
\vdots & \vdots & \vdots & \ddots & \vdots & \vdots \\
0 & 0 & 0 & \cdots & \frac{1}{\lambda} & -\frac{1}{\lambda^2} \\
0 & 0 & 0 & \cdots & 0 & \frac{1}{\lambda}
\end{pmatrix}.
\tag{3.1.3}
$$

One gets equality (3.1.1) from (3.1.2) for $f(z) = z^n$. Really, for $n > 0$ in this case

$$
\frac{f^k(\lambda)}{k!} = \frac{n(n-1)(n-2)\ldots(n-k)}{k!} \lambda^{n-k} = C_n^k \lambda^{n-k}.
$$

Equality (3.1.2) holds for the function $f(z) = z^n$ if n is negative, but in this case

$$
\frac{n(n-1)(n-2)\ldots(n-k)}{k!} = (-1)^k C_{-n}^k = (-1)^k C_{|n|}^k.
$$

In this way, J^n for negative n is given by the formula

$$
J^n = \sum_{s=0}^{m-1} (-1)^s C_{|n|}^s N^s \lambda^{n-s}.
$$

If $\sigma = \text{sign}(n)$, for arbitrary integer n we get

$$
J^n = \sum_{s=0}^{m-1} \sigma^s C_{|n|}^s N^s \lambda^{n-s}.
$$

In the matrix presentation, this equation is written in the form of

$$
J^n = \begin{pmatrix}
\lambda^n & \sigma C_{|n|}^1 \lambda^{n-1} & \sigma^2 C_{|n|}^2 \lambda^{n-2} & \cdots & \sigma^{m-1} C_{|n|}^{m-1} \lambda^{n-m+1} \\
0 & \lambda^n & \sigma C_{|n|}^1 \lambda^{n-1} & \cdots & \sigma^{m-2} C_{|n|}^{m-2} \lambda^{n-m+2} \\
0 & 0 & \lambda^n & \cdots & \sigma^{m-3} C_{|n|}^{m-3} \lambda^{n-m+3} \\
\vdots & \vdots & \vdots & \ddots & \vdots \\
0 & 0 & 0 & \cdots & \sigma C_{|n|}^1 \lambda^{n-1} \\
0 & 0 & 0 & \cdots & \lambda^n
\end{pmatrix}.
\tag{3.1.4}
$$

Note also, that the transpose matrix $J^\top$ and adjoint matrix J^* are note functions of the Jordan cell J in the previous sense. But, these matrices (in the same basis) are lower-triangular Jordan cells, and one has analogous formulas for the powers of these matrices.

For further computations, the matrix J^n is useful to express in a somewhat different form. Let $\lambda = r\omega$, where $r = |\lambda|$, $|\omega| = 1$. From the obvious expression

$$J = r(\omega I + \frac{1}{r}N) = r(U + T),$$

where matrix $U = \omega I$ is a unitary matrix, and matrix $T = \frac{1}{r}N$—nilpotent, for positive n we get

$$J^n = r^n \sum_{s=0}^{m-1} C_n^s U^{n-s} T^s = r^n U^n \sum_{s=0}^{m-1} C_n^s U^{-s} T^s.$$

Analogously, for negative n we obtain

$$J^n = r^n \sum_{s=0}^{m-1} C_{|n|}^s \sigma^s U^{n-s} (-T)^s.$$

3.2 Representations of the Powers of the Operator

The restriction of the operator A to the subspace $L(k, j, i)$ is given by the Jordan cell $J(k, j, i)$. This Jordan cell can be represented in the form

$$J(k, j, i) = \lambda(k, j)I + N(k, j, i),$$

where $N(k, j, i)$ is a nilpotent operator acting on the basis vectors by the formulas

$$N(k, j, i)e(k, j, i, 1) = 0,$$

$$N(k, j, i)e(k, j, i, l) = e(k, j, i, l - 1) \text{ for } l > 1, \qquad (3.2.5)$$

and $I = I(k, j, i)$ is the identity operator on the space $L(k, j, i)$. In the following, for the simplicity of formulas, we shall sometime omit indices k, j, i in the notation for these operators.

Let $A(k, j)$ be the restriction of the operator A to the invariant subspace $L(k, j)$. Then

$$A(k, j) = \bigoplus_i J(k, j, i) = \lambda(k, j)I + N(k, j),$$

where $I = I(k, j)$ is the identity operator on the space $L(k, j)$, and the operator

$$N(k, j) = \bigoplus_i N(k, j, i)$$

is nilpotent.

Let $A(k)$ be the restriction of the operator A to the invariant subspace $L(k)$. Then, for operator A, we have the representation of

$$A = \bigoplus_k A(k) = \bigoplus_k \bigoplus_j A(k, j) = \bigoplus_k \bigoplus_j [\lambda(k, j)I + N(k, j)].$$

Since the subspaces in question are invariant, while taking the powers of the operator, one has to take the powers of all summands appearing in the direct sum. Therefore, applying previously obtained formulas for $J(k, j, i)^n$, we obtain the expression for A^n.

The Jordan cell $J(k, j, i)$ has the form $J(k, j, i) = \lambda(k, j)I + N(k, j, i)$, where $N(k, j, i)$ is a nilpotent matrix of corresponding dimension, having numbers 1 on the second diagonal, and all other entries are 0. For $r(k) > 0$ we introduce numbers $\omega(k, j) = r(k)^{-1}\lambda(k, j)$ and present the Jordan cell in the form of

$$J(k, j, i) = r(k)[\omega(k, j)I + T(k, j, i)],$$

where $T(k, j, i) = r(k)^{-1}N(k, j, i)$. Such a presentation of the Jordan cell induces the presentation

$$A(k) = r(k)[U(k) + T(k)], \tag{3.2.6}$$

$$A(k, j) = r(k)[U(k, j) + T(k, j)],$$

where operators $U(k)$ and $U(k, j)$ are unitary, and operators $T(k)$ and $T(k, j)$—nilpotent.

Let

$$p(k, j) = \max_i q(k, j, i),$$

$$p(k) = \max_j p(k, j),$$

$$p = \max_k p(k).$$

These numbers are maximum among dimensions of Jordan cells which correspond, respectively, to eigenvalue $\lambda(k, j)$; eigenvalues with modulus $r(k)$; p is the maximum among dimensions of all Jordan cells.

Operators $T(k, j)$ and $U(k, j)$ commute. Therefore, it is possible to obtain presentations for positive powers directly, using Newton binomial formula:

$$[A(k, j)]^n = \{r(k)[U(k, j) + T(k, j)]\}^n = A(k, j)^n$$

$$= r(k)^n \sum_{s=0}^{n} C_n^s U(k, j)^{n-s} T(k, j)^s.$$

Since

$$[T(k, j)]^{p(k,j)} = 0,$$

some of the summands in the last sum vanish, and finally we obtain

$$A(k, j)^n = r(k)^n \sum_{s=0}^{p(k,j)-1} C_n^s U(k, j)^{n-s} T(k, j)^s.$$

For the operator $[A(k)]^n$, we obtain representation

$$A(k)^n = r(k)^n \sum_{j=1}^{q(k)} A(k, j)^n$$

$$= r(k)^n \sum_{j=1}^{q(k)} \sum_{s=0}^{p(k,j)-1} C_n^s U(k, j)^{n-s} T(k, j)^s.$$

This expansion may also be written in the form

$$A(k)^n = r(k)^n \sum_{s=0}^{p(k)-1} C_n^s U(k)^{n-s} T(k)^s.$$

From the above considerations, we obtain a representation of the operator A^n.

Theorem 3.2.1 *For positive powers of a nonsingular operator A one has*

$$A^n = \sum_{k=1}^{q} r(k)^n \sum_{s=0}^{p(k)-1} C_n^s U(k)^{n-s} T(k)^s \tag{3.2.7}$$

$$= \sum_{k=1}^{q} r(k)^n \sum_{j=1}^{q(k)} \sum_{s=0}^{p(k,j)-1} C_n^s U(k, j)^{n-s} T(k, j))^s,$$

where operators $U(k)$ and $U(k, j)$ are unitary, and operators $T(k)$ and $T(k, j)$— nilpotent, are given according to (3.2.6).

Applying the operator A^n to the vector $x \in L$, we obtain an explicit presentation of the vector trajectory:

$$A^n x = \sum_{k=1}^{q} r(k)^n \sum_{s=0}^{p(k)-1} C_n^s U(k)^{n-s} T(k)^s x(k), \qquad (3.2.8)$$

where $x(k) = P(k)x$.

Remark An analogous expansion holds for a singular operator A. In this case $r(1) = 0$, and in the operator expansion, one or more of the Jordan cells $J(1, 1)$ occur with $\lambda = 0$. As noted before, the formula 3.1.4 simplifies for $\lambda = 0$. Therefore, the expansion for A^n has the form

$$A^n x = N(1)^n x(1) + \sum_{k=2}^{q} r(k)^n \sum_{s=0}^{p(k)-1} C_n^s U(k)^{n-s} T(k)^s x(k)$$

$$= N(1)^n x(1) + \sum_{k=2}^{q} r(k)^n \sum_{j=1}^{q(k)} \sum_{s=0}^{p(k,j)-1} C_n^s U(k, j)^{n-s} T(k, j))^s x(k). \qquad (3.2.9)$$

At the same time, $N(1)^n x(1) = 0$ for n large enough and the first summand in the last formula does not affect the asymptotic behavior of the sequence. In the following, we always assume that the operator A in nonsingular.

3.3 The Complete Expansion of the Vector Trajectory

3.3.1 Positive Semi-trajectories

The obtained expansion (3.2.8) of the trajectory of the vector x is convenient for the study of the behavior of this trajectory, since the essential dependence on n under this expansion only occurs through numerical coefficients $r(k)^n$ and C_n^s. Indeed, the operators $U(k)$ are unitary, therefore the norm of the vector $U(k)^{n-s} T(k)^s x(k)$ does not depend on n, and the behavior of the norm of the summand in (3.2.8) depends on the numerical multiplier $r(k)^n C_n^s$ only.

For fixed s, the binomial coefficient C_n^s increases like n^s as $n \to +\infty$. Therefore, if $T(k)^s x(k) \neq 0$, the norm of the summand

$$r(k)^n C_n^s U(k)^{n-s} T(k)^s x(k)$$

behaves under the growth of n as $r(k)^n n^s \|T(k)^s x(k)\|$, and norms of all non-zero summands in (3.2.8) have different asymptotic behavior.

According to with the rate of growth of the norm when $n \to +\infty$, these summands are linearly ordered lexicographically, i.e. under the following rule:

$$r(k_1)^n C_n^{s_1} \preccurlyeq r(k_2)^n C_n^{s_2}, \quad \text{if } k_1 < k_2 \text{ and if } k_1 = k_2 \text{and } s_1 \leqslant s_2.$$

In particular, for a given k, the highest rate of growth has the non-zero summand with the highest index s. Note that this order is the same as the order (2.2.7), induced by inclusion of subspaces $M^s(k)$ from (2.2.8).

For a vector x in general position, all summands in the previous representation of the trajectory are non-zero, and the highest rate of growth has the summand with the largest k and the largest s corresponding to such k, i.e. it is the summand $C_n^{p(q)} U(q)^{n-p(q)} T(q)^{p(q)} x(q)$. But, for different vectors x, different summands may be non-zero, and the behavior of the trajectory of the concrete vector x depends on the form of these summands.

Let us describe which of the summands, appearing in the expansion (3.2.8), are non-zero for the given vector x, i.e. let us find out when $T(k)^s x(k) \neq 0$.

Let $x(k) \neq 0$. Consider the expansion

$$x(k) = \sum_{l=1}^{p(k)} Q(k; l)x(k),$$

where operator $Q(k; l)$ is the projector onto the subspace $L^l(k)$, introduced by formula (2.2.5).

Let $p_k(x)$ be the largest of the indices l, such that $Q(k; l)x(k) \neq 0$. In other words, $p_k(x)$ is the highest of the orders of adjoint vectors, actually appearing in the expansion of the vector $x(k)$. Note that the operators $Q(k; l)$ depend upon the choice of the Jordan basis. However, the introduced characteristic $p_k(x)$ of the vector x is canonical—it does not depend upon the choice of a basis. Really, it follows from the definition that

$$p_k(x) = \min\{s : x(k) \in L^s(k)\}.$$

Since the subspaces $L^s(k)$ are canonically determined, the magnitude $p_k(x)$ is also canonically determined. This magnitude may also be given as the index of the level $L^s(k) \setminus L^{s-1}(k)$ of the pyramid $L^1(k) \subset L^2(k) \subset \ldots \subset L(k)$, containing the vector $x(k)$.

Let us show that the representation

$$x(k) = \sum_{l=1}^{p_k(x)} Q(k; l)x(k)$$

of the vector $x(k)$ allows the determination of non-zero terms in the expansion (3.2.8).

For $l > 1$, the operator $T(k)$ maps the subspace $L^l(k)$ injectively into the subspace $L^l(k-1)$, and maps the subspace $L^1(k)$ to zero. Therefore, $T(k)^s x(k, l) = 0$ for $s \geq l$. In this way, $T(k)^s x(k) = 0$ for $s \geq p_k(x)$, and for $s < p_k(x)$, we obtain representation of the $T(k)^s x(k)$ in the form of the sum of $p_k(x) - s$ summands:

$$T(k)^s x(k) = T(k)^s x(k, p_k(x)) + T(k)^s x(k, p_k(x)-1) + \ldots + T(k)^s x(k, p_k(x)-s).$$
$$(3.3.10)$$

In particular, for the maximal value $s = p_k(x) - 1$, we obtain only one summand

$$T(k)^{p_k(x)-1}x(k) = T(k)^{p_k(x)-1}x(k, p_k(x)).$$

This vector is non-zero and belongs to the subspace $L^1(k)$, generated by eigenvectors of the operator $A(k)$.

For the next value $s = p_k(x) - 2$, we obtain two summands

$$T(k)^{p_k(x)-2}x(k) = T(k)^{p_k(x)-2}x(k, p_k(x)) + T(k)^{p_k(x)-2}x(k, p_k(x) - 1).$$

Here, the vector $T(k)^{p_k(x)-2}x(k, p_k(x))$ is non-zero, and it belongs to the subspace $L^2(k)$. The vector $T(k)^{p_k(x)-2}x(k, p_k(x)-1)$ belongs to the subspace $L^1(k)$. Therefore, the summands are linearly independent and the sum is a non-zero vector belonging to the subspace $M^2(k)$, given by (2.2.6).

Analogously, for $0 \le s < p_k(x)$, the summand $T(k)^s x(k, p_k(x) - s)$ from (3.3.10) belongs to the subspace L^1 and is different from zero, all summands are linearly independent, thus sum is a non-zero vector belonging to the space $M^{p_k(x)-s}(k)$.

As a result, we have obtained all non-zero terms of the expansion of the trajectory of the vector $x(k)$. Let us write these terms in decreasing order, with respect to the rate of growth of the norm:

$$
\begin{aligned}
A^n x(k) = A(k)^n x(k) &= r(k)^n C_n^{p_k(x)-1} U(k)^{n-s} T(k)^{p_k(x)} x(k, p_k(x) - 1) \\
&+ r(k)^n C_n^{p_k(x)-2} U(k)^{n-p_k(x)-1} [T(k)^{p_k(x)-1} x(k, p_k(x) - 1) \\
&+ T(k)^{p_k(x)-1} x(k, p_k(x) - 2)] \\
&+ r(k)^n C_n^{p_k(x)-3} U(k)^{n-p_k(x)-2} [T(k)^{p_k(x)-2} x(k, p_k(x) - 1) \\
&+ T(k)^{p_k(x)-2} x(k, p_k(x) - 2) + T(k)^{p_k(x)-2} x(k, p_k(x) - 2)] + \ldots \\
&+ r(k)^n U^n x(k).
\end{aligned}
$$

For the description of the trajectory of an arbitrary vector x, we use the following characteristics of a vector x:

$$
\begin{aligned}
K(x) &= \{k : x(k) \ne 0\}, \\
p_k(x) &= \min\{s : x(k) \in L^s(k)\}, \\
k(x) &= \max K(x), \\
s(x) &= p_{k(x)}(x).
\end{aligned}
$$

These are the same numbers introduced before—the pair of numbers $(k(x), s(x))$ is the index of the level of the thickened pyramid (2.2.8), containing the vector x.

Note that the number $k(x)$ is the largest of values k, thus at least one of the coordinates $x(k, j, i, l)$ is different from zero. The number $s(x)$ is the largest of the values l, thus at least one of the coordinates $x(k(x), j, i, l)$ is different from zero.

This discussion proves the following statement concerning the complete expansion of the vector trajectory.

Theorem 3.3.1 (*On the complete expansion of the positive semi-trajectory of a vector*). *Let A be an invertible operator of a finite-dimensional vector space. For the elements of the trajectory of the vector x, one has the following expansion*

$$A^n x = \sum_{k \in K(x)} \sum_{s=0}^{p_k(x)-1} r(k)^n C_n^s U(k)^{n-s} T(k)^s x(k), \qquad (3.3.11)$$

which uses magnitudes and operators introduced before.

All the terms in this expansion are non-zero, they all have different asymptotic behavior as $n \to +\infty$:

$$\| r(k)^n C_n^{p_k(x)-l} U(k)^{n-p_k(x)-l} T(k)^{p_k(x)-l} x(k) \| \sim r(k)^n n^{p_k(x)-l},$$

and they are naturally ordered in according to the order relation (2.2.7).

The first terms of the expansion, ordered according to the decreasing rate of growth of norm, have the form

$$A^n x = r(k(x))^n C_n^{s(x)-1} U(k(x))^{n-s(x)} T(k(x))^{s(x)-1} x(k(x))$$
$$+ r(k(x)))^n C_n^{s(x)-2} U(k(x))^{n-s(x)+1} T(k(x))^{s(x)-2} x(k(x)) + \dots$$
$$+ r(k(x))^n U(k(x))^n x(k(x)) + \dots.$$

In particular, the principal term of the expansion of the trajectory is the sequence

$$w_n = r(k(x))^n C_n^{s(x)-1} U(k(x))^{n-s(x)} T(k(x))^{s(x)-1} x(k(x)),$$

and this sequence has the form

$$y_n = r(k(x))^n C_n^{s(x)-1} U^n(w),$$

where the vector
$$w = U(k(x))^{-s(x)} T(k(x))^{s(x)-1} x(k(x))$$

belongs to the subspace $M(k(x); 1)$, i.e. it is a linear combination of eigenvectors of the operator A with eigenvalues $\lambda(k(x), j)$.

Remark This construction of an asymptotic expansion uses a scale of the sequences of the form $r^n C_n^s$, indexed by two indices and naturally ordered. The more standard in analogous situations is the expansion using exponential-power double scale, with functions of the form $r^n n^s$, and instead of the constructed expansion, one may pass to the expansion by the scale $r^n n^s$.

However, in the problem under investigation, the expansion by the scale $r^n C_n^s$ is more convenient. Firstly, it appears naturally and secondly, this expansion is finite and contains simple enough expressions for the terms of the expansion. If one uses the expansion by the scale $r^n n^s$, one obtains expansion containing infinitely many terms and, additionally, the expressions for the coefficients in such expansion are more complex.

3.3.2 Negative Semi-trajectories

According to (3.1.4), the expression for the matrix J^{-n} in the original bases has practically the same form as for positive powers (only the signs in some of the terms are different). Therefore, for negative n, we obtain completely analogous expression for the operator A^n, an analogous form of the vector trajectory, and the form of nonzero terms in the trajectory expansion.

The main difference here is that under the transition to the inverse matrix, the order of numeration of moduli of eigenvalues changes, and when $n \to -\infty$ the rate of growth of the sequences, generates different order:

$$r(k_1)^n C_n^{s_1} \precsim r(k_2)^n C_n^{s_2}, \text{ if } k_1 > k_2 \text{ and if}$$
$$k_1 = k_2 \text{ and at the same time } s_1 \leqslant s_2. \tag{3.3.12}$$

Therefore, the principal term of the expansion is the summand with index $k = k'(x)$, where $k'(x) = \min\{k : x(k) \neq 0\}$. It may happen that $k'(x) \neq k(x)$, but it is a true fact $k'(x) \leq k(x)$. The dependence on s remains the same: for a given k, the highest rate of growth has the nonzero summand with the highest index s. Thus, for the characterization of the principal term in the expansion of the negative semi-trajectory of the vector x, we introduce the pair of numbers $(k'(x), s'(x))$, where $k'(x)$ is the smallest of all values k where at least one of the coordinates $x(k, j, i, l)$ is different from zero, and $s'(x)$ is the largest of all values l, where at least one of the coordinates $x(k'(x), j, i, l)$ is different from zero.

Theorem 3.3.2 (*On the complete expansion of the negative vector semi-trajectory*). *Let A be an invertible operator in a finite-dimensional space. For the terms of the negative semi-trajectory of the vector x, one has the expansion*

$$A^{-n}x = \sum_{k \in K(x)} \sum_{s=0}^{p_k(x)-1} \frac{1}{r(k)^n}(-1)^s C_n^s U(k)^{-n-s} T(k)^s x(k)$$

which uses previously defined values and operators.

In this expansion, all the terms are nonzero, have different asymptotic behavior as $n \to +\infty$, and they are naturally ordered according to the order relation (3.3.12).

In particular, the principal term of the expansion is the sequence

$$w_n' = \frac{1}{r(k'(x))^n} C_n^{s'(x)-1} U(k'(x))^{-n-s'(x)} T(k'(x))^{s'(x)-1} x(k'(x)),$$

where this sequence belongs to the subspace $L^1(k'(x))$, generated by eigenvectors of the operator A corresponding to the eigenvalues $\lambda(k'(x), j)$.

References

1. Godunov, B., & Zabreiko, P. (1995). Geometric characteristics for convergence and asymptotics of successive approximations of equations with smooth operators. *Studia Mathematica, 116*, 225–238.
2. Gantmakher, F. R. (1988). *Theory of atrices*. Moscow: Nauka.

Chapter 4
The Structures Related to the Principal Term of the Vector Trajectory

In this section, we present an analysis of the formula (3.3.11), and concentrate our attention to the relation between the obtained expansion of the trajectory and some well known mathematical notions. Such an analysis is useful for the comprehension of the specifics of the problems under consideration, and the place of the emerging structures among general mathematical structures. In addition to that, the facts that we get by this analysis, may serve as good examples, useful in the study of the corresponding notions, for example, of ultrametric spaces.

Let us consider the principal term of the expansion (3.3.11) for given x:

$$y_n = r(k(x))^n C_n^{s(x)-1} U(k(x))^{n-s(x)} T(k(x))^{s(x)-1} x(k(x)).$$

This sequence may also be represented in the form

$$y_n = r(k(x))^n C_n^{s(x)-1} U(k(x))^n (w),$$

where the operator $U(k(x))$ is unitary, and the vector

$$w = U(k(x))^{s(x)} T(k(x))^{s(x)-1} x(k(x))$$

does not depend on n.

So, in the expression (3.3.11) of the principal term the following objects appear:

1. The map χ, which associates to the vector x the number $r(k(x))$ and, consequently, the sequence $r(k(x))^n$;
2. The map ψ, which associates to the vector x the pair of numbers $(k(x), s(x))$ and consequently, the sequence $r(k(x))^n C_n^{s(x)-1}$;
3. The map W, which associates to the vector x the vector

$$w = W(x) = U(k(x))^{-s(x)} T(k(x))^{s(x)-1} x(k(x)),$$

giving the direction of the principal term of the trajectory asymptotic.

© Springer International Publishing Switzerland 2014
Ć.B. Dolićanin and A.B. Antonevich, *Dynamical Systems Generated by Linear Maps*,
DOI: 10.1007/978-3-319-08228-8_4

4.1 The Lyapunov Exponent and the Local Spectral Radius

First of all, let us note that the map χ is a special case of the Lyapunov exponent [1].

An (abstract) exponent on a m-dimensional vector space X is a function $\chi(x)$ with values in $\mathbb{R}$, defined everywhere on X, except on the zero element, having the properties:

(1) $\chi(cx) = \chi(x)$ for $c \neq 0$;
(2) $\chi(x + y) \leq \max\{\chi(x), \chi(y)\}$.

It is sometimes convenient to put $\chi(0) = -\infty$ and then these properties hold for all points in the space. The term *indicator* for the exponent can be used too.

This concept arose in the theory of systems of differential equations, where indicators appear as characteristics of the rate of growth of the solutions of the system.

The exponent of growth (more precisely, the upper of characteristic exponent) of the real-valued function $x(t)$ is the number

$$\chi(x) = \overline{\lim_{t \to +\infty}} \frac{1}{t} \ln |x(t)|.$$

Analogously, *the exponent of growth* for the sequence of real numbers x_n is the number

$$\chi(x) = \overline{\lim_{n \to +\infty}} \frac{1}{n} \ln |x_n|.$$

For functions (or sequences) with values in a normed space, the indicator of growth is defined as the exponent of growth of the norm.

The exponent of growth of the vector-function $x(t)$ with values in a normed space X is the magnitude

$$\overline{\lambda} = \chi(x) = \overline{\lim_{t \to +\infty}} \frac{1}{t} \ln \|x(t)\|.$$

If the space X is finite-dimensional, it is irrelevant which norm is used on the space of values, since in a finite-dimensional space all norms are equivalent, and the indicator of growth does not depend on the choice of a norm. If λ is a finite exponent of the vector-function $x(t)$, then for each $\varepsilon > 0$, for t large enough, one has

$$\|x(t)\| \leq Ce^{(\lambda+\varepsilon)t},$$

and there is a sequence $t_n \to +\infty$ such that

$$\|x(t_n)\| > e^{(\lambda-\varepsilon)t_n}.$$

The reverse is also true—a number λ having these properties is the exponent of that vector-function. Thus, the exponent of growth characterizes the exponential rate of growth of that function.

The fundamental objects used for exponents in the investigation are solutions of linear systems of differential equations. The set of solutions of the linear system of m differential equations

$$\frac{du}{dt} = A(t)u(t),$$

where $A(t)$ is a matrix-valued function, form an m-dimensional space. This space may be parameterized, for example, by initial conditions $u(0) = x$, lying in a finite-dimensional space X. The exponents of growth of the solutions of the system is a function on X, having the properties (1) and (2), i.e. this function is an abstract exponent in the aforementioned sense.

The fact that the exponents arose as some characteristics of solutions of systems of differential equations have no part in the various topics under consideration. The basic properties of exponents are consequences of axioms (1) and (2).

We have discussed, previously, a pyramid (2.2.3) of subspaces $M(k)$ and the number $k(x)$ was given as the index of that level of the pyramid, containing the given vector. This is a manifestation of a general pattern: for arbitrary exponent there exists a pyramid of subspaces associated to this exponent.

Theorem 4.1.1 ([1], Chap. 1, p. 2) *A function χ on a finite-dimensional subspace is an exponent if and only if there is a pyramid*

$$(0 = L_0 \subset L_1 \subset L_2 \subset \cdots \subset L_p)$$

such that χ has constant values λ_j on the levels S_j and these values distributed according to the increasing order:

$$-\infty \leq \lambda_0 < \lambda_1 < \ldots, < \lambda_p < +\infty.$$

Proof Suppose that a function χ has constant values on the levels of the pyramid and

$$-\infty \leq \lambda_0 < \lambda_1 < \ldots, < \lambda_p < +\infty.$$

Let us consider two arbitrary vectors x and y; for definiteness, we assume that $\chi(x) = \lambda_j \leq \chi(y) = \lambda_k$. Then $x \in S_j \subset L_j, y \in S_k \subset Ł_k$. Since the elements of the pyramid are linearly ordered according to the growth χ, we have $L_j \subseteq L_k$, where $x + y \in L_k$ and $\chi(x + y) \leq \chi(y) = \max\{\chi(x), \chi(y)\}$.

Further, on if $x \in S_j$, then for the constant $c \neq 0$ we have $cx \in S_j$, where we get $\chi(cx) = \chi(x)$.

Now, suppose that χ is an exponent and let λ_j be one of the values of that function. Consider the set

$$L_j = \{x \mid \chi(x) \le \lambda_j\}.$$

For arbitrary vectors x and y from L_j, we have

$$\chi(x + y) \le \max\{\chi(x), \chi(y)\} \le \lambda_j,$$

and we obtain that $x + y \in L_j$. We also have $\chi(cx) = \chi(x) \le \lambda_j$, where $cx \in L_j$. This means that the set L_j is a vector subspace. In additionally, it is obvious that if $\lambda_k < \lambda_j$ then $L_k \subset L_j$.

So, for the exponent $\chi(x)$ a corresponding pyramid is constructed.   □

The weight of the level S_j is called *the multiplicity* of the corresponding value λ_j of the exponent $\chi(x)$.

The correspondence between exponents and pyramids is, obviously, not one-one: to each exponent, there is one corresponding pyramid, but on a pyramid, one may give infinitely different exponents. In particular, if f is an arbitrary strictly increasing function, and χ is an exponent on, then $f(\chi)$ is also the exponent corresponding to the same pyramid. For example, one may take the function $f(t) = e^t$. In particular, in the following problems, it is sometimes convenient to use the positive exponent $e^{\chi(x)}$ in place of an exponent $\chi(x)$.

The most important properties of the exponent are determined only by the structure of the corresponding pyramid.

So, the function

$$\chi : x \mapsto r(k(x)), \tag{4.1.1}$$

naturally appearing under the construction of the expansion of the vector trajectory, is an exponent and the corresponding pyramid is the pyramid (2.2.3).

In the theory of linear operators, in considering sequences of the form $A^n x$ instead of the exponent of the growth, one usually considers the local spectral radius of the vector.

Let us recall that *the spectral radius* of a bounded linear operator A in a Banach space is the number

$$R(A) = \max\{|\lambda| : \lambda \in \sigma(A)\}.$$

In particular, for an operator A of a finite-dimensional vector space, the spectral radius is equal to the maximum of all moduli of eigenvalues, i.e. to the number r_q.

For the spectral radius, one has the Berling-Gelfand formula:

$$R(A) = \lim_{n \to \infty} \|A^n\|^{1/n}.$$

The local spectral radius of the vector x (with a given operator A) is the number

$$R(x) = \varlimsup_{n \to \infty} \|A^n x\|^{1/n}.$$

Directly from the expansion (3.3.1), we obtain

Corollary 4.1.2 *For the given linear operator A in $\mathbb{C}^m$, the local spectral radius of the vector x is given by the formula*

$$R(x) = r(k(x)).$$

The local spectral radius, as a function of the variable x, is an exponent and the pyramid corresponding to it is the pyramid (2.2.3).

The results obtained may serve to illustrate the following known result.

Proposition 4.1.3 *Let A be a bounded linear operator of the Banach space X. For an arbitrary vector $x \in X$, one has an estimate*

$$R(x) \leq R(A), \tag{4.1.2}$$

where for almost all the vectors, one has equality (the set of all vectors for which $R(x) < R(A)$ is a set of the first category).

In the example of an operator on a finite-dimensional vector space, we have an explicit description of the set on which in (4.1.2) one has equality.

Corollary 4.1.4 *For the given linear operator A in $\mathbb{C}^m$, the equality*

$$R(x) = R(A)$$

holds for all x from the last level of the pyramid (2.2.3)—the set

$$S_q = M(q) \setminus M(q-1),$$

and the strict inequality holds only for vectors from the set $M(q-1)$, which is a subspace of smaller dimension, it is nowhere dense, and, in particular, is a set of the first category in $\mathbb{C}^m$.

We shall also discuss what is the relation between the exponent $\chi(x) = r(k(x))$, induced by the sequence A^n and the exponent $\chi_{-1}(x) = r(k'(x))$, induced by the sequence A^{-n}.

There, the pyramid (2.2.3) corresponding to $\chi(x)$, has the form

$$M(k) = \bigoplus_{s=1}^{k} L(s),$$

but to the exponent $\chi_{-1}(x)$ corresponds another pyramid, having the form

$$M'(k) = \bigoplus_{s=k}^{q} L(s).$$

Let us list some facts from the theory of exponents that we shall need in the future. One usually performs computations by using the given basis and it is natural to choose the basis according to the properties of an exponent, more precisely, in accordance to the structure of the corresponding pyramid.

Let an exponent $\chi(x)$ be given on a space X, and let $(v_1, v_2, \dots, v_m)$ be an arbitrary basis of X. The values $\chi(v_k)$ are called *basis exponents*. If a basis is indexed in such a way that $\chi(v_k) \leq \chi(v_{k+1})$, this basis is called *ordered*.

A basis $(v_1, v_2, \dots, v_m)$ is called *normal* with respect to the pyramid L, if the number of the basis vectors belonging to the level is equal to the weight of that level.

A linear combination of some vectors u_j, with non-zero coefficients c_j is called *lowering* if

$$\chi \left(\sum c_j \, u_j \right) < \max\{\chi(u_j)\}.$$

We shall say that a pyramid is *complete*, if this pyramid is not a part of any other pyramid. Obviously, the pyramid is complete if and only if all levels have weight 1. The corresponding exponent takes m different finite values. Such an exponent shall be called *complete*.

Proposition 4.1.5 *Let*

$$L = (L_1 \subset L_2 \subset \cdots \subset L_p)$$

be a pyramid and $\chi(x)$ be an arbitrary exponent generating this pyramid.

The following properties of a basis (v_j) are equivalent:

1. Basis is normal, i.e. the number of basis vectors belonging to any level is equal to the weight of that level.

2. The number of basis vectors belonging to the subspace L_k is equal to the dimension of that subspace.

3. The sum of the exponents of basis vector, i.e. the number

$$\sum_{j=1}^{m} \chi(v_j),$$

is minimal. In particular, for the case of a complete exponent, this property holds, if and only if, the exponents of all basis vectors are different.

4. No subset of basis vectors has any lowering combination.

For a normal basis, the most natural is the numeration under this basis is ordered.

A construction of normal bases is one of the most basic technical methods in the theory of the Lyapunov exponents. A. M. Lyapunov used the following process of transformation of an arbitrary basis into a normal one. We list the corresponding assertion, together with a proof, since it is a basis for a part of the subsequent construction.

Theorem 4.1.6 (Lyapunov) *Let one has a pyramid in a space X, and let $\chi(x)$ be an arbitrary exponent corresponding to that pyramid. For any basis $(v_1, v_2, \ldots, v_m)$, there is such a nonsingular triangular matrix C so that the new basis $(u_1, u_2, \ldots, u_m)$, where $u_j = Cv_j$, i.e.,*

$$u_1 = c_{11}v_1,$$

$$u_2 = c_{12}v_1 + c_{22}v_2,$$

$$\ldots$$

$$u_m = c_{1m}v_1 + c_{2m}v_2 + \cdots + c_{mm}v_m,$$

is normal for a given pyramid. At the same time, $\chi(u_j) \leq \chi(v_j)$ holds.

Corollary 4.1.7 *There is a normal basis for any pyramid (and, consequently, for any exponent).*

Proof Let us conduct the process of transformation of the basis (v_j) to a normal one, in the following way. Let $c_{kk} = 1$ and, consequently, we look for u_k in the form

$$u_k = c_{1k}v_1 + c_{2k}v_2 + \cdots + c_{k-1,k}v_{k-1} + v_k.$$

For any k, choose coefficients, where the indicator of the vector u_k is the smallest. Due to the finiteness of the set of values of indicators, such a linear combination always exists. If there are several of such combinations, we choose any of them.

Let us show that the collection of vectors we get in this way is, indeed, a normal basis. In order to do that, let us check that no part of this collection of vectors allow any lowering combinations. From the proposition 2, follows that this collection is a normal basis.

Suppose that a lowering combination exists, i.e. for some nonzero numbers b_{jk} one has

$$\chi(u) < \max \chi(u_{jk}),$$

where

$$u = \sum_i b_{jk} u_{jk}. \tag{4.1.3}$$

Let us split the sum (4.1.3) into two groups of summands. We distinguish, in the sum, the leading terms, i.e. these summands where the indicator achieves maximal values and separately list lower terms, having smaller indicators. Since the sum of lower terms has the indicator strictly smaller than the number $\max \chi(u_{jk})$, and the whole sum also has the indicator smaller than that number, it follows that the sum of leading terms also has indicator strictly smaller than the number $\max \chi(u_{jk})$.

So, if for some collection of vectors u_j there is a lowering linear combination, then the lowering linear combination also exists for the leading terms, appearing in the given collection. Therefore, we may assume that all the vectors u_{j_k}, in the formula (4.1.3), have the same indicator, i.e. all vectors are leading onces.

Let k_0 be the largest of indices j_k, appearing in (4.1.3). Since all numbers b_{j_k} are nonzero, we may assume, without the loss of generality, that $b_k = 1$. Replacing the vectors u_{j_k} for $j_k < k_0$ by their expressions, we obtain a linear combination of the form (4.1.3), where the indicator is smaller than the indicator of the vector u_{k_0}. But, by the construction of vectors u_{k_0}, this vector is a linear combination with the smallest indicator. We arrive to a contradiction. There, no part of the collection of vectors u_j allows any lowering combinations, and this collection is a normal basis with respect to the given pyramid.   □

Note that the numeration of the normal basis obtained in the theorem is not ordered.

4.2 Non-archimedean Norms

One may look at the properties of indicators from a different point of view if one pays the attention to the connection between indicators and the theory of non-archimedean norms on vector spaces and the theory of ultrametric spaces.

A metric space (X, ρ) is *ultrametric* if the metric is ultrametric—the following axiom, stronger that the triangle inequality, holds:

$$\varrho(x, y) \leq \max\{\varrho(x, z), \varrho(z, y)\}.$$

In textbooks, one usually gives ultrametric spaces as examples of metric spaces with exotic geometry. In such a space, for example, any two balls of the same radius are either disjoint, or they are identical. Therefore, for any $r > 0$, the space is a disjoint union of balls of radius r. Each of these balls, (if it is not a set containing only one point), is also a disjoint union of balls of smaller radius etc. The following property also looks strange—any point in a ball is its center.

The geometry of ultrametric spaces fundamentally differs from the usual geometry of the Euclidean spaces, and our geometric intuition, which is formed in the process of investigation of the Euclidean spaces, does not work here.

Ultrametric naturally arises in a number of problems, in particular in p-adic analysis, and in connection to non-archimedean norms. Let us show that these structures naturally arise in the asymptotic theory, and that the facts, exotic in comparison with the geometry of the Euclidean spaces, are quite natural and obvious in the context of asymptotic theory.

Let us give the necessary definitions.

A norm on a field K is a map $\varphi : K \to \mathbb{R}$, satisfying conditions

(i) $\varphi(\lambda) \geq 0$;

(ii) $\varphi(\lambda + \mu) \leq \varphi(\lambda) + \varphi(\mu)$;

(iii) $\varphi(\lambda\mu) = \varphi(\lambda)\varphi(\mu)$;

(iv) $\varphi(\lambda) = 0 \Leftrightarrow \lambda = 0$.

In the case of number fields $\mathbb{Q}$, $\mathbb{R}$ and $\mathbb{C}$, an example of a norm is the modulus of a number, and these conditions are a record of the basic properties of the modulus. Therefore, instead of $\varphi(\lambda)$, it is common to use the notation $|\lambda|_K$ or $|\lambda|$. In these fields, there are other norms. An example is *the trivial norm* $|\lambda|_0$, which may be given in any field by:

$|\lambda|_0 = 1$ for $\lambda \neq 0$ and $|0|_0 = 0$.

The particular attention was paid to the investigation of norms in the field $\mathbb{Q}$ of rational numbers. In the field $\mathbb{Q}$, for every prime number p, there is the so-called p-adic norm $|\lambda|_p$, which is given as follows. Let p be a prime number. For the rational number λ, there is its prime factorization (with integer, including negative, exponents) $\lambda = p^\nu p_1^{\nu_1} \ldots p_k^{\nu_k}$, where p_i are some other prime numbers. One then puts $|\lambda|_p = p^{-\nu}$. The axioms may easily be verified.

In the p-adic theory, one of the fundamental facts is the Ostrowski theorem, which claims that in the field $\mathbb{Q}$, there are no other norms, except those listed: the usual modulus $|\lambda|$, trivial norm $|\lambda|_0$ and p-adic norm $|\lambda|_p$.

A norm may be archimedean and non-archimedean. These notions arise it in relation to the Archimedean axiom in the theory of real numbers, which may be written in the following form.

The Archimedean axiom. *If* $x \neq 0$ *and* $|x| < |y|$, *then there is a natural number* n, *such that* $|nx| > |y|$.

A norm $|\cdot|$ on a field is called *archimedean*, if this axiom holds.

A norm $|\cdot|$ on a field is called *non-archimedean*, if from the condition $|x| < |y|$ it follows that $|nx| < |y|$ for any natural number n.

Lemma 4.2.1 *Let* $|\cdot|$ *be a norm in a field* K. *The following conditions are equivalent:*

(1) *the stronger triangle inequality holds*

$$|x + y| \leq \max\{|x|, |y|\}.$$

(2) *it is non-archimedean norm;*

(3) *the set of natural numbers is bounded;*

(4) *the set of all natural numbers is bounded by the number 1:* $|n| \leq 1$ *for any* n.

Proof (1) $\Rightarrow$ (2). Let $|x| < |y|$. Then $|2x| = |x + x| \leq |x|$, $|3x| = |2x + x| \leq |x|$, and for any n, one has $|nx| \leq |x| < |y|$, i.e. the norm is non-archimedean.

(2) $\Rightarrow$ (3). Let $|x| < |y|$. Then for each n, one has $|nx| = |n||x| \leq |y|$. From this, one concludes that the set of numbers $|n|$ is bounded.

(3)$\Rightarrow$(4). If for some n_0, we have $|n_0| > 1$, then $|n_0^k| = |n_0|^k \to \infty$ and this contradicts (3).

(4) $\Rightarrow$ (1). From the triangle inequality, we have

$$|x + y|^n = |(x + y)^n| = |\sum_{k=0}^{n} C_n^k x^k y^{n-k}| \leq \sum_{k=0}^{n} |C_n^k x^k y^{n-k}| \leq$$

$$\leq \sum_{k=0}^{n} |x|^k |y|^{n-k} \leq (n+1) \max\{|x|^n, |y|^n\}.$$

Therefore,

$$|x+y| \leq \lim_{n\to\infty} \sqrt[n]{n+1} \; \max\{|x|, |y|\} = \max\{|x|, |y|\}.$$

$\square$

The usual modulus (on the fields $\mathbb{Q}$, $\mathbb{R}$ and $\mathbb{C}$) is an archimedean norm, the trivial norm is non-archimedean on any field, all p-adic norms on $\mathbb{Q}$ are non-archimedean.

A *norm* on a vector space X (over the normed field K with the norm $|\cdot|_K$), is a map $\|\cdot\| : X \to \mathbb{R}$, satisfying the conditions:

(Ni) $\|x\| \geq 0$, $\|x\| = 0 \Leftrightarrow x = 0$;

(Nii) $\|\lambda x\| = |\lambda|_K \|x\|$;

(Niii) $\|x+y\| \leq \|x\| + \|y\|$.

A normed space is a metric space, if one introduces a metric by the standard formula $\varrho(x, y) = \|x - y\|$.

A norm on a vector space is *non-archimedean*, if the stronger triangle inequality holds

$$\|x+y\| \leq \max\{\|x\|, \|y\|\}.$$

In particular, the metric given by the non-archimedean norm is an ultrametric, and under the non-archimedean norm, one deals with unusual geometry on a vector space.

The notion of a norm is a generalization of the length of a vector, and for any norm the number $\|x\|$, characterizes the magnitude of the vector x. In the case of the non-archimedean norm, we obtain that if $\|x\| < \|y\|$, then $\|nx\| < \|y\|$ for all $n \in \mathbb{N}$. This property can be interpreted as follows: if $\|x\| < \|y\|$, then the element x is infinitely small with respect to y. In such a way, one may interpret the fact that the Archimedean axiom holds as the condition in the absence of infinitely small elements different from 0, and non-archimedean norms naturally occur in the cases, when some elements are infinitely small, with respect to others.

The asymptotic theory has to deal with the situation where some elements are infinitely small with respect to others, and the relation to non-archimedean norms naturally occurs in this theory.

Let us consider objects appearing during the investigation of vector trajectories, using the introduced notions.

First of all, note that the axioms of the Lyapunov exponents are almost the same as the axioms for a norm on a vector space, with the trivial norm on the field over. The difference is in the fact that the values of exponents may be negative. But, this difference is not essential. For an arbitrary exponent $\chi(x)$, one may construct an exponent taking only nonnegative values, for example, one may consider the exponent $e^{\chi(x)}$. After that, one may look at the Lyapunov exponent as a special case of the non-archimedean norm.

In our case, the local spectral radius $R(x) = r(k(x))$, induced by the operator A, determines the non-archimedean norm on $\mathbb{C}^m$, with the trivial norm on the field $\mathbb{C}$.

Let us show, in the form of an example, how exotic properties of the geometry of ultrametric spaces (and spaces with non-archimedean norm) may be interpreted in the language of the asymptotic theory or the linear algebra, and they become natural and evident from this point of view.

In the example, the non-archimedean norm (the number $R(x) = r(k(x))$), characterizes the exponential rate of growth of the sequence $\|A^n x\|$. Therefore, for example, the equality $\|\lambda x\| = \|x\|$, reflects the fact that the rate of growth of the sequence does not change under the multiplication of the sequence with the nonzero number. A stronger triangle inequality also reflects the following simple fact—it shows that the rate of growth of the sum of two sequence may not be larger than the rate of growth of the summands.

The ball $B[0, r]$ (of radius r with the center at the point 0) consists of those points x, whose exponential rate of growth of the trajectory does not exceed r; this ball is a vector subspace M_r. The sphere $S[0, r]$ (with the center at the point 0 and radius r) consists of those points x, whose exponential rate of growth of the trajectory is r; as follows from the previous discussion, such a sphere is one of the levels of the pyramid corresponding to the indicator under investigation—the local spectral radius.

The ball $B[x_0, r]$ (of radius r with the center at the point x_0) consists of those points x, whose rate of growth of the trajectory of the difference $x - x_0$ does not exceed r, i.e. $x \in B[x_0, r]$ if and only if $x - x_0 \in M_r$. In the language of linear algebra, this means that each ball is an equivalence class for the relation: $x \asymp y$ if $x - y \in M_r$.

It is, therefore, obvious, that equivalence classes are mutually disjoint: in the language of metric spaces, this means that an ultrametric space splits into disjoint balls of fixed radius. All elements from an equivalence class are of the same importance—geometrically, this corresponds to the fact that any point in a ball is its center: for every $x \in B[x_0, r]$ we have $B[x, r] = B[x_0, r]$.

Let us consider which result corresponds to the fact from the theory of ultrametric spaces that each ball splits into balls of smaller radius. In the example under consideration, this ball $B[0, r_j]$ is a vector subspace L_j, a ball of radius $r_i < r_j$ is a vector subspace $L_i \subset L_j$. The subspace L_j splits into equivalence classes of the equivalence relation $x \sim y \Leftrightarrow x - y \in L_i$; additionally, each equivalence class is a ball of radius r_i.

It is possible to note other properties of the space with non-archimedean norm, evident from the point of view of the asymptotic theory, and exotic in comparison to the geometry of the Euclidean spaces. For example, let us consider the ball $B[x_0, r]$, under the condition that $r < \|x_0\|$. The elements of that ball are the points of the form $x_0 + z$, where the rate of growth of the trajectory of the point z does not exceed r, and is strictly smaller than the rate of growth of the trajectory of the point x_0. But, then, the rate of growth of each of this quantities is the same as the rate of growth of the trajectory of the point x_0, i.e. $\|x_0 + z\| = \|x_0\|$. This fact, from the geometrical point of view, means that under the condition $r < \|x_0\|$, the ball $B[x_0, r]$ is contained in the sphere

$$S(0, \|x_0\|) = \{x : \|x\| = \|x_0\|\}.$$

These remarks are not formally used in the text that follows, but they clarify the notions of the local spectral radius, and, in fact, these features have to be considered in subsequent calculations.

4.3 The Adjusted Indicator and Generalized Non-archimedean Norms

Let us now move to the consideration of a map, which describes more detailed the rate of growth of the sequence $\|A^n x\|$. Namely, let us consider a pair of numbers $(k(x), s(x))$, corresponding to a vector x, defined in such a way that the sequence $\|A^n x\|$ has the same behavior as the sequence $r(k(x))^n n^{s(x)}$. Here, we encounter a situation which does not fit in the scheme of the theory of exponents, and the theory of non-archimedean normed spaces, since one corresponds to a vector a pair of numbers, not just one number. In this case, a generalization of the notion of non-archimedean norm naturally arises.

The following observation may serve as a basis for this generalization. First of all, let us note that the pair of number $(k(x), s(x))$, is an interim technical notation; essentially we correspond to any vector x one of the sequence of the form $r^n n^s$, and in the discussion, the most important is only the fact that the set of such sequences is a linearly ordered set. Let us also note, that in the definition of an ultrametric space, (and in the case of non-archimedean norm on the vector space with the trivial norm on a number field), the axioms use the relation $\leq$, for the values of the norm $\|x\|$, and use the fact that for two values $\|x\|$ and $\|y\|$ the quantity $\max\{\|x\|, \|y\|\}$ is determined. Here, the operations of addition and multiplication are not used. But, the relation $\leq$ and the notion of maximum of two elements is determined for elements of any linearly ordered set. Therefore, it is possible to modify the definition of the non-archimedean norm, using instead of $\mathbb{R}$ any linearly ordered set.

Let X be a vector space over a field K, with the trivial norm, and Λ be any linearly ordered set which has minimal element 0. *A generalized norm* on X is a map $\| \cdot \| : X \to \Lambda$, which has the properties

(Ni) $\|x\| \geq 0$, $\|x\| = 0 \Leftrightarrow x = 0$;

(Nii) $\|\lambda x\| = \|x\|$ for $\lambda \neq 0$;

(Niii) $\|x + y\| \leq \max\{\|x\|, \|y\|\}$.

In this way, the same axioms as in the classical case on non-archimedean norms, distinguish the class of spaces with the new structure.

The analogous generalization of the notion of an ultrametric space naturally arises.

A *generalized ultrametric* on the set X with values in a linearly ordered set Λ is a map $\varrho : X \times X \to \Lambda$ where the same axioms hold as for the usual definition of an ultrametric. An generalized ultrametric space is a set with a given generalized ultrametric. Under this definition, the basic properties of ultrametric space are saved.

Note that different generalizations of the notion of a norm were considered in relation to other problems. For example, in nonstandard analysis one considers norms taking values in nonstandard extension $\tilde{\mathbb{R}}$ of the field $\mathbb{R}$, being a linearly ordered set (and it is a field). In fact, in our situation, each sequence $r^n s^n$ determines a nonstandard number, and the generalized norm under consideration may be seen as the special case of a norm, with values in $\tilde{\mathbb{R}}$. But, nonstandard analysis is oriented toward more complex situations, and we shall not use the terminology from nonstandard analysis here.

In the given problem, concerning the behavior of the trajectory $A^n x$, the set $\Lambda = \mathbb{R}_0^+ \times \mathbb{N}_0$ naturally occurs and Λ is linearly ordered lexicographically:

$$(r, s) \prec (r_1, s_1) \text{ if } r < r_1 \text{ and if } r = r_1, s \le s_1. \tag{4.3.4}$$

And the generalized non-archimedean norm appears: the map

$$\chi_1 : x \to (r(k(x)), s(x)) \in \mathbb{R}_0^+ \times \mathbb{N}_0, \tag{4.3.5}$$

which we shall call *the adjusted indicator* or *the adjusted norm*.

The proofs of the listed propositions pass to the case of a norm with values in a linearly ordered set. In particular, with the adjusted indicator (4.3.5), the thickened pyramid (2.2.8) is related, and it allows more detailed description of the properties of the trajectory $A^n x$.

4.4 Graded-Linear Operators

Let us consider the map $W : x \to w$, associating to a vector x the vector w giving the direction of the principal term in the asymptotic expansion of the trajectory:

$$W : x \to w = U(k(x))^{-s(x)} T(k(x))^{s(x)-1} x(k(x)). \tag{4.4.6}$$

This map acts by the formula

$$\begin{aligned} w = W(x) &= U(k(x))^{-s(x)} T(k(x))^{s(x)-1} x(k(x)) \\ &= U(k(x))^{-s(x)} T(k(x))^{s(x)-1} P(k)x, \end{aligned}$$

i.e. one obtains w from x as the result of the application of the linear operator $U(k(x))^{-s(x)} T(k(x))^{s(x)-1}$ to the projection $x(k(x))$ of the vector x. But, the map W is not linear, since for different points x, one has to use different linear operators.

The form of the corresponding linear operator depends on the pair of numbers $(k(x), s(x))$, i.e. it is determined by the value of the generalized norm.

The vectors having the same norm (or generalized norm), from the geometric point of view, form the sphere. But, under the non-archimedean norm, the balls with the center at the point 0 are subspaces, and they form a pyramid, while the spheres with the center at the point 0 are levels of that pyramid. This means we deal with the map which is given by different linear operators on different levels of the pyramid.

In our case, the pyramid corresponding to the generalized norm

$$\|x\| = (k(x), s(x)),$$

is the pyramid (2.2.8), consisting of the previously defined subspaces $M^s(k)$, and the action of the map W is given by the formula

$$W(x) = U(k)^{-s}T(k))^{s-1}P(k)x, \quad \text{if} \ \ x \in M^s(k) \setminus M^{s-1}(k).$$

Let us list some properties of the map W.

This map is homogeneous ($W(tx) = tW(x)$).

For arbitrary x and z we have $A^n(x+z) = A^n x + A^n z$. If $\|z\| < \|x\|$, then $\|A^n z\|$ is a quantity, infinitely small with respect to $\|A^n x\|$. Therefore, the principal term in the expansion of the trajectory $A^n(x+z)$ is the same as the principal term in the expansion of the trajectory $A^n x$.

Hence, we obtain the property: if $\|z\| < \|x\|$, then $W(x+z) = W(x)$.

In particular, W is not a linear mapping.

If $\|z\| < \|x\|$, then $\|x+z\| = \|x\|$, and element $x+z$ belongs to the same level $M^s(k) \setminus M^{s-1}(k)$ of the pyramid. On this level, the map W acts as a linear operator, so from equality $W(x+z) = W(x)$ it follow that the vector z belongs to the kernel of the corresponding linear operator. Here the corresponding linear operator is $U(k)^{-s}T(k)^{s-1}P(k)$ and this property can be werify directly.

If $x \in M^s(k) \setminus M^{s-1}(k)$, we have $W(x) \in M^1(k)$ and, consequently, mapping W is order-bounded: for the generalized norm one has

$$\|W(x)\| \le \|x\|, \tag{4.4.7}$$

where the inequality is understood in the sense of the order (4.3.4). This inequality looks as the condition that the operator is bounded, but W is not continuous in the natural topology.

In this way, in the problem on asymptotic behavior of the trajectory of a vector, the following class of discontinuous maps naturally occurs, which we shall call graded-linear.

Let on a vector space X is given the generalized non-archimedean norm $\|\cdot\|$, and let $\{M_j\}$ pyramid corresponding to that norm.

Definition 4.4.1 A map $F : X \to X$ is called graded-linear operator, if there exists a pyramid $\{M_j\}$ and exist linear operators $T_j : M_j \to X$, such that
(i) $F(x) = T_j x$ for $x \in M_j \setminus M_{j-1}$;
(ii) $T_j x = 0$ for $x \in M_{j-1}$.
  A graded-linear operator shall be called order-bounded if

$$\|F(x)\| \leq \|x\|.$$

Graded-linear operators inherit some of the properties of linear operators, but there are many differences. Among them, let us note that the image of a linear operator is always a vector subspace, while the image of a graded-linear operator is a union of vector subspaces.

In this way, the map W, giving the principal term in the expansion of the trajectory, is a order-bounded graded-linear map. The operators giving other terms in the expansion have an analogous structure.

It is precisely the appearance of these operators, (nonlinear and discontinuous in the natural topology of the space), being characteristic for the asymptotic problems, and they determine their specificity. During the investigation of the behavior of the trajectory of vectors, this appears naturally, and may be directly investigated. But, during the investigation of trajectory of subspaces, one has to deal with such maps, and the described structure is essentially used in the calculation in the part II, and therefore, we focus our attention to peculiar properties of these maps.

1. We have the simplest example in the Perron case, when the operator A has m different eigenvalues, and these eigenvalues have different moduli. Then the non-archimedean norms (4.1.1) and (4.3.5) are the same, and are given by the expression $\|x\| = r(k(x))$, where $k(x)$, is the largest of indices of nonzero coordinates of the vector x.

The graded-linear operator W acts by the following rule: to the vector

$$x = (x(1), x(2), \ldots, x(k), 0, \ldots, 0), x(k) \neq 0,$$

one associates the vector

$$w = (0, 0, \ldots, x(k), 0, \ldots, 0).$$

Such an operator W, acts on the vector x as the projector to the one-dimensional subspace $L(k(x))$, generated by the vector $e(k(x))$. In this example, the image of the operator W is the union of one-dimensional subspaces $L(k)$.

2. Let the matrix of linear operator A consists of two Jordan cells of dimension 2, and let $|\lambda_1| < |\lambda_2|$. Then, the corresponding graded-linear operator W acts on the vector $x = (x_1, x_2, x_3, x_4)$ by the following rule

$$Wx = \begin{cases} (0, 0, \frac{1}{\lambda_2}x_4, 0), & x_4 \neq 0, \\ (0, 0, \frac{1}{\lambda_2}x_3, 0), & x_4 = 0, x_3 \neq 0, \\ (\frac{1}{\lambda_1}x_2, 0, 0, 0), & x_4 = x_3 = 0, x_2 \neq 0, \\ (\frac{1}{\lambda_1}x_1, 0, 0, 0), & x_4 = x_3 = x_2 = 0, x_1 \neq 0. \end{cases}$$

Reference

1. Bylov, B. F., Vinograd, R. E., Grobman, D. M., Nemyckii, V. V. (1966). *Lyapunov exponents.* Moscow: Nauka.

Chapter 5
The Asymptotic Behavior of Vector Trajectories and Trajectories of One-Dimensional Subspaces

Detailed description of vector trajectories and one-dimensional subspaces conducts have been given in this chapter. Conducts of mentioned trajectories have been determined by the main member of complete development of vector trajectories. The influence of unitary operator and linear operator at sphere has been discussed.

5.1 The Behavior of the Vector Trajectory

The expansion (3.3.1) allows a detailed description of the behavior of the vector trajectories, and trajectories of one-dimensional subspaces. In these cases, the behavior of these trajectories is determined solely by the principal term in the expansion. As it shall shown below, if $\dim V > 1$, then for the description of the behavior of the trajectory of subspaces V one may require all other terms of the asymptotic expansion.

Let us first discuss the vector trajectories. We shall use previously introduced notation, in particular, we shall use the indicator (4.1.1), the adjusted indicator (4.3.5), and the local spectral radius $R(x) = r(k(x))$.

The condition $r(x) = 1$, $s(x) = 1$, from the following theorem, express the fact the vector $w = P(k(x))x$ is a linear combination of eigenvectors of the operator A, corresponding to the eigenvalues with modul 1.

Theorem 5.1.1 *Let A be an arbitrary linear invertible operator in the space X.*

1. $A^n x \to 0$ *if and only if $R(x) < 1$.*
2. $\|A^n x\| \to \infty$ *if and only if one of the following conditions hold:*
(2i) $R(x) > 1$

or

(2ii) $R(x) = 1$ *and $s(x) > 1$.*
3. *If $R(x) = 1$ and $s(x) = 1$, then the trajectory $A^n x$ is bounded and approaches to the trajectory of the vector*

© Springer International Publishing Switzerland 2014
Ć.B. Dolićanin and A.B. Antonevich, *Dynamical Systems Generated by Linear Maps*,
DOI: 10.1007/978-3-319-08228-8_5

$$w = P(k(x))x,$$

i.e. to the sequence

$$U(k(x))^n w.$$

4. *The trajectory $A^n x$ has nonzero limit if and only if the corresponding vector w is an eigenvector of the operator A with eigenvalue 1. Then $A^n x \to w$.*

The proof follows directly from the given expansion.

For a more detailed description of the trajectory $A^n x$, let us discuss behavior of the sequence of norms $\|A^n x\|$. The behavior of the sequence of directional vectors $\frac{1}{\|A^n x\|} A^n x$ shall be discussed later.

Let us introduce the notation for the relation between numerical sequences a_n and b_n:

$$a_n \sim b_n, \quad \text{if } a_n = O(b_n) \text{ and } b_n = O(a_n),$$

i.e. if the sequences $\frac{a_n}{b_n}$ and $\frac{b_n}{a_n}$ are bounded;

$$a_n \approx b_n, \quad \text{if } \frac{a_n}{b_n} \to 1.$$

The equality

$$\lim_{n \to \infty} \|A^n x\|^{1/n} = r(k(x))$$

obtained before, characterizes behavior of the norm $\|A^n x\|$ up to the subexponential factor. From the form of the trajectory, one may obtain more detailed information on the behavior of this sequence, in particular, describe the adjusted order of the growth of the sequence $\|A^n x\|$, and obtain the explicit value of the numerical coefficient for the sequence from the scale $r^n n^s$.

Theorem 5.1.2 *The sequence of norms $\|A^n x\|$ have the following asymptotic behavior:*

$$\|A^n x\| \approx r(k(x))^n n^{s(x)-1} \frac{1}{(s(x)-1)!} \|w\|,$$

In particular,

$$\|A^n x\| \sim r(k(x))^n n^{s(x)-1}.$$

Proof The behavior of the sequence of norms is determined by the principal term y_n of the expansion of the trajectory, where

$$y_n = r(k(x))^n C_n^{s(x)-1} U(k(x))^n (w),$$

the operator $U(k(x))$ is unitary and so

$$\|y_n\| = r(k(x))^n C_n^{s(x)-1} \|w\|.$$

Since,

$$C_n^s \approx \frac{n^s}{s!},$$

we obtain the given description of the asymptotic behavior of the norm, using a more standard scale. $\qquad\square$

5.2 The Action of a Unitary Operator

If $r(x) = 1$ and $\nu(x) = 1$, then, according to the claim 3) of the Theorem 5.1.1, the question on behavior of the vector trajectory is reduced to the question on the behavior of the trajectory of vectors $w = P(k(x))x$. These vectors belong to an invariant subspace, where the operator A is unitary. Therefore, let us discuss the question concerning the behavior of the trajectory under the action of an arbitrary unitary operator.

Let U be a unitary operator in a finite-dimensional space. For such an operator, there is a basis of eigenvectors, and moduli of eigenvalues are 1. Therefore, without a loss in generality, we may assume the space under consideration is $\mathbb{C}^d$, and that the operator is given by a diagonal matrix:

$$U = diag\{\omega_1, \omega_2, \ldots, \omega_d\}, \quad \text{where } \omega_k = e^{i2\pi h_k}, \ h_k \in \mathbb{R}.$$

Under the action of such an operator, the trajectory of the vector x converges if and only if this vector is an eigenvector with eigenvalue 1.

Let us describe the form of the set $\Omega(x)$ of limit points of the trajectory for arbitrary x. In our case, $\Omega(x)$ is the closure of the trajectory of the point x. By definition, this is the minimal closed subset containing the trajectory of the point x, and $\Omega(x)$ is invariant with respect to the action of the operator. In order to describe such sets, we construct splittings of the space $\mathbb{C}^d$ in the invariant closed subsets.

This can be done as follows. Consider the map

$$\varphi(x) = (|x_1|, |x_2|, \ldots, |x_d|). \tag{5.2.1}$$

This map acts from $\mathbb{C}^d$ to $\mathbb{R}^d$, and the image is the closed cone of positive vectors $\mathbb{R}^d_+$. Since obviously $\varphi(Ux) = \varphi(x)$, the preimage of each point from $\mathbb{R}^d_+$ is a nonempty closed invariant set.

The preimage of a point $\xi \in \mathbb{R}_+^d$ has the form

$$\varphi^{-1}(\xi) = \{x \in \mathbb{C}^d : |x_k| = \xi_k\}.$$

It is obvious that if $\xi_k = 0$, then $x_k = 0$, and if $\xi_k \neq 0$, then $x(k) = z_k \xi_k$, where $|z_k| = 1$.

In this way, the points from $\varphi^{-1}(\xi)$ may be naturally parameterized by the collection of numbers z_k with indices k, for which $\xi_k \neq 0$, and satisfying the condition $|z_k| = 1$. This means that the set $\varphi^{-1}(\xi)$ is homeomorphic to the product of finitely many circles, i.e. it is a torus, which we denote by $\mathbf{T}_\xi$. The dimension of that torus is equal to the number of nonzero coordinates of the vector ξ, and we denote that dimension by $\nu(\xi)$.

In particular, for interior points of the set $\mathbb{R}_+^d$, all coordinates are nonzero and the dimension of the torus $\mathbf{T}_\xi$ is d; for boundary points of $\mathbb{R}_+^d$ corresponding tori have dimensions from 1 to $d - 1$. To the point 0, corresponds the degenerate torus consisting of only one point.

Any torus

$$\mathbf{T}^\nu = \{z = (z_1, z_2, \ldots, z_\nu), \ |z_k| = 1\}$$

is a group with respect to the operation of coordinate-wise multiplication, the identity element is $(1, 1, \ldots, 1)$. Each torus T_ξ is invariant with respect to the action of the operator U, where the action of the map U on a torus is given as coordinate-wise multiplication by the vector $\omega(\xi)$, formed by the numbers ω_k with index k, for which $\xi_k \neq 0$. Such a map of torus into itself is called *standard shift of a torus*.

The behavior of trajectories of points on a torus, under the action of the standard shift is well known. From that behavior the solution of a series of known problems depend, for example, the problem of small denominators, etc. The foundational results were obtained by Weyl [1].

On a torus $\mathbf{T}^\nu$, let us discuss the trajectory of the identity element under the action of the standard shift induced by the element $\omega = (\omega_1, \omega_2, \ldots, \omega_\nu) \in \mathbf{T}^\nu$, i.e. the set $\{\omega^n, n \in \mathbb{Z}\}$, consisting of powers of the element ω. Let us denote H_ω the closure of that trajectory. Then H_ω is a closed subgroup of the torus $\mathbf{T}^\nu$.

By using this subgroup, we obtain the description of the closure of the trajectory for an arbitrary point $z \in \mathbf{T}^\nu$. The trajectory of the point z is the set $\{\omega^n z, n \in \mathbb{Z}\}$; this set belongs to the equivalence class $[z]$ of the point z in the quotient group $\mathbf{T}^\nu / H_\omega$. By the construction of the subgroup H_ω the trajectory $\{\omega^n z, n \in \mathbb{Z}\}$ dense in that equivalence class is every where and the class $[z]$ is a closed set. From these considerations, one has to describe the set $\Omega(z)$.

Proposition 5.2.1 *The closure of the trajectory of the point $z \in \mathbf{T}^\nu$, under the action of the standard shift, generated by the element*

$$\omega = (\omega_1, \omega_2, \ldots, \omega_\nu) \in \mathbf{T}^\nu, \omega_k = e^{i 2\pi h_k},$$

is the equivalence class of that point in the quotient group T^ν / H_ω:

$$\Omega(z) = \overline{\{\omega^n z, n \in \mathbb{Z}\}} = [z] \in T^\nu / H_\omega. \qquad \square$$

This proposition reduces the investigation of the trajectory to the description of the subgroup H_ω. The analysis shows that the form of this subgroup depends very much on the arithmetic properties of numbers h_k, which determine the element ω.

In this problem, the notion of the rational independence of a collection of real numbers, is essential.

The numbers $h_1, h_2, \ldots, h_n$ are called *rationally independent*, if from the equality

$$q_1 h_1 + q_2 h_2 + \ldots + q_d h_n = 0,$$

where the coefficients are integers (equivalently rational numbers), it follows that all $q_k = 0$.

For example, the rational independence of numbers 1 and h is equivalent to the condition that the number h is irrational.

The rational independence of two irrational numbers h_1 and h_2 is equivalent to the condition that the number h_1 / h_2 is irrational.

If the numbers h_1 and h_2 are irrational, (rationally independent with 1), and mutually rationally independent, from this one still cannot be concluded that the triple $1, h_1, h_2$ is rationally independent, since it may be possible to obtain the relation of the form $q_1 h_1 + q_2 h_2 = q_3$.

Here one has to note that it is possible to consider $\mathbb{R}$ as a vector space over field of rational number $\mathbb{Q}$. From that point of view, the definition of the rational independence of numbers is the usual definition of linear independence of h_k, as elements of that vector space over $\mathbb{Q}$. This space is infinite dimensional, in particular for each n there exists a collection of n rationally independent numbers. As an example, one may look at the numbers of the form $\sqrt{p}$, where p is a prime number. But, for arbitrary concrete irrational numbers checking whether these numbers are rationally independent it is a difficult problem. For example, it is not known at this moment, whether the numbers e, π and 1 are rationally independent.

The form of the subgroup H_ω (and consequently, the behavior of the trajectory of points on a torus under the action of the standard shift), depends on the existence of rational relations between number $1, h_1, h_2, \ldots, h_\nu$ and, in particular, depends on the number $d(\omega)$ being rationally independent among the numbers $1, h_1, h_2, \ldots, h_\nu$. For different ω, this number may take the values from 1 to $\nu + 1$. The simplest form of the subgroup H_ω may be described in two extreme cases.

Suppose that all numbers h_k are rational, and let N be the smallest common denominators of these numbers. Then, the subgroup H_ω is finite, and it is isomorphic to the cyclic finite group $\mathbb{Z}_N$. In this case, the trajectory of each point is periodic with period N. This is the case when $d(\omega) = 1$ and for numbers $1, h_1, \ldots, h_\nu$ there are ν independent linear relations, with integer coefficients.

In the opposite case, when $d(\omega) = \nu + 1$, i.e. the numbers $1, h_1, \ldots, h_\nu$ are rationally independent, the set of points ω^n is dense in the torus everywhere, $H_\omega = \mathbf{T}^\nu$ and the trajectory of each point is dense everywhere in the torus $\mathbf{T}^\nu$. Note that in the books on dynamical systems, exactly this case is most thoroughly discussed, since only in this case the standard shift is an *ergodic* map.

If $1 < d(\omega) < \nu$, then there are nontrivial rational relations among the numbers $1, h_1, \ldots, h_\nu$. Among these numbers, there are irrational, and they may be some rational numbers. In this case, the set of points ω^n is not dense everywhere and the form of the subgroup H_ω has a more complicated description. Depending on the form of rational relations, the subgroup H_ω is isomorphic either to the torus $\mathbf{T}^{d(\omega)}$ of dimension $d(\omega)$, either to the product $\mathbf{T}^{d(\omega)} \times \mathbb{Z}_N$ of that torus with a finite cyclic subgroup.

This reasoning proves the following theorem.

Theorem 5.2.2 *Under the action of the unitary operator U, given by a diagonal matrix, the space $\mathbb{C}^d$ stratifies by the map*

$$\varphi(x) = (|x_1|, |x_2|, \ldots, |x_d|)$$

into invariant tori $\mathbf{T}_\xi$, parameterized by the points ξ from the closed positive cone $\mathbb{R}_+^d$. In each torus, the action of the operator U is the same as the action of the standard shift, induced by the element $\omega(\xi)$. The set $\Omega(x)$ of limit points of the trajectory is the closure of the trajectory of the point x, and it is the equivalence class $[x]$ of that point in the quotient group $\mathbf{T}_{\varphi(x)}/H_{\omega(\varphi(x))}$.

Note that, if there are eigenvalues equal to 1, then among tori $\mathbf{T}_\xi$ there are those all of whose points are fixed.

In particular, this theorem gives more detailed description of the behavior of the trajectory of points x, for which $r(x) = 1$ and $s(x) = 1$. Since the trajectory $A^n x$ of such a point tends toward the trajectory of the vector w, the set of limit points of the trajectory $A^n x$ is equal to the set of limit point of the trajectory of the point w. The point w belongs to a subspace where the operator acts as unitary. From this, we get

Theorem 5.2.3 *If $r(x) = 1$ and $s(x) = 1$, then the trajectory $A^n x$ tends toward the trajectory of the vector w and*

$$\Omega(x) = \Omega(w) = [w] \in \mathbf{T}_{\varphi(w)}/H_{\omega(\varphi(w))}.$$

5.3 The Action of a Linear Operator on a Sphere

In the description of the trajectory of the point, it is natural to follow not only the behavior of the points of the trajectory $A^n x$, but also the behavior of the "directional vectors" $\frac{1}{\|A^n x\|} A^n x$. These vectors lie on a unit sphere, and in fact the question reduces to the description of the trajectory of the map of the sphere $S^{2m-1} \subset \mathbb{C}^m$ into itself, given by the formula

$$\alpha(x) = \frac{1}{\|Ax\|} Ax, \; x \in S^{2m-1}.$$

This map naturally arises in a number of other problems, and it is important to describe the dynamics of such a map.

Note that a point x from the sphere induces a ray (the set of the form $L_x = \{tx : t > 0\}$), where the correspondence between the points on the sphere and on the ray is bijective. Therefore, this problem is also the problem concerning description of the behavior of rays, under the action of linear maps.

Theorem 5.3.1 *For each vector x, the sequence of directional vectors $\frac{1}{\|A^n x\|} A^n x$ tends toward the sequence $\frac{1}{\|A^n w\|} A^n w$ of directional vectors of the trajectory of the vector w. If $s(x) > 1$, then one has an estimate of the rate of convergence*

$$\|\frac{1}{\|A^n x\|} A^n x - \frac{1}{\|A^n w\|} A^n w\| = O(1/n).$$

If $s(x) = 1$, then the rate of convergence is exponential:

$$\|\frac{1}{\|A^n x\|} A^n x - \frac{1}{\|A^n w\|} A^n w\| = O([\frac{r(k(x) - 1)}{r(k(x))}]^n n^{q_0}),$$

where

$$q_0 = \max_{j,i} q(k_0 - 1, j, i),$$

i.e. it is the maximal dimension of the Jordan sell corresponding to $r(k(x) - 1)$.

Proof We obtain a proof by a direct analysis of the form of the trajectory. The trajectory has the form

$$A^n x = y_n + \tau_n,$$

where y_n is the principal term in the expansion of the trajectory of the vector x, given by the formula

$$y_n = r(k(x))^n C_n^{s(x)-1} U(k(x))^n(w) = C_n^{s(x)-1} A^n w.$$

In the general case, we have the following estimation for the remainder

$$\|\tau_n\| = O(\frac{1}{n}[r(k(x))]^n C_n^{s(x)-1}).$$

If $s(x) = 1$, then the norm of the second term of the expansion is estimated, using powers of the number $r(k(x) - 1)$, which is smaller than $r(k(x))$:

$$\|\tau_n\| = O([r(k(x) - 1)]^n n^{q_0}).$$

Therefore, the quotient exponentially decreases and

$$\frac{\|\tau_n\|}{\|y_n\|} = O([\frac{r(k(x) - 1)}{r(k(x))}]^n n^{q_0}).$$

Since

$$\frac{1}{\|A^n w\|} A^n w = \frac{1}{\|y_n\|} y_n,$$

we get

$$\|\frac{1}{\|A^n x\|} A^n x - \frac{1}{\|A^n w\|} A^n w\| = \|[\frac{1}{\|A^n x\|} - \frac{1}{\|y_n\|}]y_n + \frac{1}{\|A^n x\|}\tau_n\|$$
$$\leq Const[(\frac{1}{\|A^n x\|} - \frac{1}{\|y_n\|}) + \frac{\|\tau_n\|}{\|A^n x\|}].$$

And for these quantities, one obtains estimations from the previous discussion.   □

Let us apply this theorem for the description of the behavior of the trajectory $\alpha^n(x)$, of the points on a sphere. Such a trajectory, according to the theorem, tends toward the trajectory of the point $\frac{1}{\|w\|} w$. We shall find the part of the sphere, where corresponding points w and their trajectories lie.

Each vector w is a linear combination of eigenvectors, corresponding to eigenvalues of equal moduli, i.e. $w \in L^1(k)$ for some k. Since this vector is of norm 1, we have $w \in L^1(k) \cap S^{2m-1}$. The dimension of the subspace $L^1(k)$ is $q(k)$, the intersection $S_k = L^1(k) \cap S^{2m-1}$ is a sphere of dimension $2q(k) - 1$, embedded in the sphere S^{2m-1}, where the spheres S_k are disjoint for different k.

In this way, on the sphere S^{2m-1} is chosen a finite collection of subsets, being spheres of smaller dimension, invariant with respect to the action of the map α such that the trajectory of the point x tends to the sphere $S_{k(x)}$ and tends toward the trajectory of the point $w \in S_{k(x)}$.

Let us discuss the action of the map α on S_k. This action is given by a unitary operator: $\alpha(w) = U(k)w$. Therefore, the problem reduces to the discussion the behavior of the trajectory under the action of the unitary operator, where this operator in the given basis of eigenvectors, already has the diagonal form. We obtain a description of the trajectory of the point under the action of the map α, on a sphere.

Let map $\varphi_k : S_k \to \mathbb{S}_+^{q(k)-1}$ is given by expression

$$\varphi_k(x) = \{|x(k, j, i, 1)|; \ 1 \leqslant j \leqslant q(k), \ 1 \leqslant i \leqslant q(k, j)\}. \tag{5.3.2}$$

Theorem 5.3.2 *The trajectory of the point $x \in S^{2m-1}$, under the action of the map α tends toward the trajectory of the point w, belonging to a sphere $S_{k(x)} \subset S^{2m-1}$.*

The sphere S_k stratifies by the map φ_k into tori and the action of the map α on each of these tori coincides with the standard shift, given by an element $\omega(w)$, induced by the corresponding numbers $\omega(k, j) = \frac{1}{r_k}\lambda(k, j)$. In relation to that, the behavior of the trajectory of the point w is given by Theorem 5.2.2

In particular, the set of limit points of the trajectory of the point x is the equivalence class $[w]$, in the corresponding quotient group $T^\nu(w)/H_\omega(w)$.

To clarify the difference between the behavior of the trajectory $A^n x$, let us discuss the behavior of the trajectory of the unit eigenvector $e(k, j)$, corresponding to the eigenvalue $\lambda(k, j)$.

Under the action of the linear map, the trajectory of the point $e(k, j)$ has the form $\lambda(k, j)^n e(k, j)$; if $|\lambda(k, j)| < 1$, then this trajectory converges to the point 0, if $|\lambda(k, j)| > 1$, this trajectory tends to infinity. Only if $|\lambda(k, j)| = 1$, then the sphere S_k is invariant and the trajectory of the point $e(k, j)$ lying on some one-dimensional tori (a circle), in particular, the vector $e(k, j)$ is a fixed point, if and only if $\lambda(k, j) = 1$.

The dynamics of the map α is completely different. The whole family of invariant spheres S_k of lower dimension appear where the action of the map is given by the corresponding unitary operator. In particular, for the map α, all eigenvectors with positive eigenvalues are fixed points.

It is possible to describe, for each sphere S_k, the *pool* $B(S_k)$—the set of points whose trajectories tends to S_k.

Theorem 5.3.3 *The pool $B(S_k)$ of the set S_k consists of points, belonging to levels with index k, of the pyramid (2.2.3), i.e.*

$$B(S_k) = S^{2m-1} \cap \{M(k) \setminus M(k-1)\}.$$

5.4 The Action of a Linear Operator on a Projective Space

The projective space $\mathbb{C}P^{m-1}$ is, by definition, the set of all one-dimensional subspaces of $\mathbb{C}^m$. Under a nonsingular linear map, a one-dimensional subspaces goes to a one-dimensional subspace. Therefore, a linear operator A induces a map

$$\beta : \mathbb{C}P^{m-1} \to \mathbb{C}P^{m-1}.$$

Such a map is called a *projective transformation*.

If a one-dimensional space L is generated by a vector x, then the subspace $A^n(L)$ is given by the vector $A^n x$. But, the behavior of the trajectory of one-dimensional space is different from the behavior of the trajectory of a vector, and different from the behavior of the trajectory under the map

$$\alpha(x) = \frac{1}{\|Ax\|} Ax.$$

For example, fixed points of the map A are, by definition, eigenvectors with eigenvalues 1, fixed points of the map

$$\alpha(x) = \frac{1}{\|Ax\|} Ax$$

are eigenvectors with positive eigenvalues, and a fixed point of the map β is any one-dimensional subspace L, spanned by an eigenvector.

Let us write a projective transformation in local coordinates. This shall help in determining how much the action of a projective transformation essentially differs from the action of a linear one.

Let us look at those vectors x, where their coordinate with a given index is nonzero. Let W_m be the set of one-dimensional subspaces, spanned by vector, whose coordinate with index m is nonzero (we use here the standard numbering of elements of a basis in $\mathbb{C}^m$). One may then, among the vectors determining a one-dimensional subspace L, uniquely choose a vector x such that $x_m = 1$. This means that the collection of numbers $(x_1, x_2, \dots, x_{m-1})$ gives local coordinates in $W_m \subset \mathbb{C}P^{m-1}$.

Let us write the action of the map β in these coordinates.

Let L be a one-dimensional subspaces with coordinates $(x_1, x_2, \dots, x_{m-1})$, i.e. this is a subspace spanned by the vector $x = (x_1, x_2, \dots, x_{m-1}, 1)$. Consider an arbitrary nonsingular matrix

$$A = \begin{pmatrix} a_{11} & a_{12} & \cdots & a_{1,m-1} & a_{1m} \\ a_{21} & a_{22} & \cdots & a_{2,m-1} & a_{2m} \\ \cdots & \cdots & \cdots & \cdots & \cdots \\ a_{m-1,1} & a_{m-1,2} & \cdots & a_{m-1,m-1} & a_{m-1,m} \\ a_{m1} & a_{m2} & \cdots & a_{m,m-1} & a_{mm} \end{pmatrix}.$$

The coordinates of the vector $y = Ax$ are given by the formula

$$y_k = \sum_{j=1}^{m} a_{kj} x_j = \sum_{j=1}^{m-1} a_{kj} x_j + a_{k,m}, \quad k = 1, 2, \dots, m.$$

In particular,

$$y_m = \sum_{j=1}^{m-1} a_{mj} x_j + a_{mm}.$$

In order to obtain coordinates of the subspace $A(L)$, in the given coordinate system, one has to divide coordinates of the vector $y = Ax$ by y_m, and then we get

$$\beta(x_1, x_2, \ldots, x_{m-1}) = \frac{1}{\displaystyle\sum_{j=1}^{m-1} a_{mj}x_j + a_{mm}} \times (\sum_{j=1}^{m-1} a_{1j}x_j + a_{1,m},$$

$$\sum_{j=1}^{m-1} a_{2j}x_j + a_{2,m} \ldots, \sum_{j=1}^{m-1} a_{m-1,j}x_j + a_{k,m}).$$

This map may be written in a more convenient matrix form.
We define the matrix $\widetilde{A}$ of format $(m-1) \times (m-1)$

$$\widetilde{A} = \begin{pmatrix} a_{11} & a_{12} & \ldots & a_{1,m-1} \\ a_{21} & a_{22} & \ldots & a_{2,m-1} \\ \ldots & \ldots & \ldots & \ldots \\ a_{m-1,1} & a_{m-1,2} & \ldots & a_{m-1,m-1} \end{pmatrix}.$$

Also, we define a column vector

$$A_m = (a_{1,m}, a_{2,m}, \ldots, a_{m-1,m})^\top,$$

composed by the elements of the last column of the matrix A, and the row vector

$$A^{(m)} = (a_{m,1}, a_{m,2}, \ldots, a_{m,m-1}),$$

composed by the elements of the last row of the matrix A. Then, assuming that x is the column vector $x = (x_1, x_2, \ldots, x_{m-1})^\top$, the expression can be written in the form of multidimensional fractional-linear map:

$$\beta(x) = \frac{1}{A_m x + a_{mm}}(\widetilde{A}x + A^{(m)}).$$

The mapping β is given by this formula only under the condition that the numerator $A_m x + a_{mm}$ is nonzero. If $A_m x + a_{mm} = 0$, then $\beta(x)$ has to be given in another coordinate system.

Example 5.4.1 Consider the one-dimensional complex projective space $\mathbb{C}P^1$, consisting of one-dimensional subspaces in the space $\mathbb{C}^2$. The vector $(z_1, z_2) \neq 0$ determines a one-dimensional subspace L, and vector (tz_1, tz_2) for $t \neq 0$ determines the same subspace. Therefore, if $z_2 \neq 0$, then the subspace L is determined by one complex number $z = \frac{z_1}{z_2}$. All vectors of the form $(z_1, 0)$, determine exactly one-dimensional subspace, where it is natural to associate the value $z = \infty$. This reasoning shows that the space $\mathbb{C}P^1$ is arranged as the extended complex plane. The extended complex plane, is, as a topological space, homeomorphic to the sphere S^2, and it is called the Riemann sphere. From the above, it follows that the linear nonsingular operator A in $\mathbb{C}^2$ given by the matrix

$$A = \begin{pmatrix} a_{11} & a_{12} \\ a_{21} & a_{22} \end{pmatrix},$$

induces on the Riemann sphere $\mathbb{C}P^1$ an ordinary fractional-linear transformation:

$$\beta(z) = \frac{a_{11}z + a_{12}}{a_{21}z + a_{22}}.$$

Note that if the denominator of the last fraction is zero, then the point z maps into the point ∞.

Now we are describing the behavior of one-dimensional subspaces under the action of a linear operator A in $\mathbb{C}^m$, i.e. the dynamics of the map β on $\mathbb{C}P^{m-1}$. Such a description is obtained from the description of the dynamics of the induced map α, on the sphere S^{2m-1}. Since proportional vectors determine the same subspace, one has to pass to the discussion of the quotient space under the equivalence relation $x \sim y$, if $x = ty, t \neq 0, t \in \mathbb{C}$. The maps being of our interest, are compatible with the structure of the quotient space—an equivalence class maps to an equivalence class. This means that $\beta([x]) = [\alpha(x)]$.

Since the trajectory of the vector $x \in S^{2m-1}$, for the action of α, tends to the trajectory of the above-constructed vector $w = Wx$, the trajectory of one-dimensional subspace L, generated by x, tends to the trajectory of one-dimensional subspace generated by w. By the homogeneity of the operator W, a vector proportional to x, transforms into a vector proportional to w, i.e. the map is compatible with the structure of the quotient space—an equivalence classes maps into an equivalence class. From this, we get that

$$\beta_n([x]) \rightarrow \beta_n([w]) = [\alpha_n(w)],$$

i.e. the trajectory of the subspace L, generated by x, tends to the trajectory of the element $[w]$.

By Theorem 5.3.2, the set of points w on the sphere splits into q subsets, each subset being a sphere of corresponding dimension, where the collection of these spheres is an attracting set—each trajectory of a point tend to that set. The factor space of the sphere S_k, for a given equivalence relation, is a projective space of the corresponding dimension.

In this way, in the projective space $\mathbb{C}p^{m-1}$, q subsets P_k were selected, where P_k is the projective space $\mathbb{C}P^{q(k)-1}$ of smaller dimension, and the trajectory of the subspace, spanned by a vector x, tends toward the trajectory of the point $[w] \in P(k(x))$. Therefore, it remains to describe the behavior of the trajectory of the equivalence class $[w]$.

By the construction, described in Theorem 5.2.2, the set of points $w \in S_k$ stratifies into tori $\mathbf{T}_\xi$ by a map of the form (5.2.1). It is immediately clear for this map to be compatible with the equivalence relation: for $x \in \mathbf{T}_\xi$, all the equivalent points also belong to $\mathbf{T}_\xi$.

Therefore, it is necessary to describe the set of equivalence classes $T_\xi/\sim$ of points of the torus, and find out how the mapping β acts on $T_\xi/\sim$.

Let us consider the torus of dimension ν:

$$T_\xi = T^q = \{w = (w_1, w_2, \ldots, w_\nu) : w_j \in \mathbb{C}, |w_k| = 1\}.$$

The equivalence class of a point w is determined by the collection of points $z_j = \frac{w_j}{w_\nu} : j = 1, 2, \ldots, \nu - 1$. Since $|z_j| = 1$, these points from $T_\xi/\sim$ are parameterized by all such collections of complex number, i.e. by the points of the torus $T^{\nu-1}$ of dimension $\nu - 1$. This correspondence is continuous and $T_\xi/\sim$, as a topological group, is the torus $T^{\nu-1}$. In the case $\nu = 1$, when the torus T_ξ is a circle, $T_\xi/\sim$ reduces to a single point.

One may look from a different point of view at the previous reasoning. Here the set of elements from T_ξ, equivalent to the unit of the group, has the form $\{(t, t, \ldots, t) : |t| = 1\}$ and this set is a (one-dimensional closed) subgroup of the group T_ξ, the so-called diagonal subgroup D. Therefore, in fact, the set $T_\xi/\sim$ is the quotient group T_ξ/D.

On the torus T_ξ, the map α acts as the standard shift:

$$\alpha(x) = (\omega_1 x_1, \omega_2 x_2, \ldots, \omega_\nu x_q), \quad \text{where } \omega_j = e^{i2\pi h_j}.$$

Therefore, on the torus $T_\xi/\sim$, the map β maps the point $z = (z_1, z_2, \ldots, z_{\nu-1})$ to the point

$$\beta(z) = (\frac{\omega_1}{\omega_\nu} z_1, \frac{\omega_2}{\omega_\nu} z_2, \ldots, \frac{\omega_{\nu-1}}{\omega_\nu} z_{\nu-1}),$$

i.e. this map acts like a standard shift on the torus of smaller dimension, generated by the element

$$\widetilde{\omega} = (\frac{\omega_1}{\omega_\nu}, \frac{\omega_2}{\omega_\nu}, \ldots, \frac{\omega_{\nu-1}}{\omega_\nu}).$$

This allows us to apply Theorem 5.2.2 to describe the behavior of the trajectory under the action of the mapping β, on each of the projective spaces P_k. Since,

$$\widetilde{\omega}_j = \frac{\omega_j}{\omega_q} = e^{i2\pi(h_j - h_q)},$$

the form of the corresponding subgroup $\widetilde{H}(\widetilde{\omega})$, generated by the element $\widetilde{\omega}$, depends on the rational relations between numbers

$$1, \widetilde{h}_1 = h_1 - h_q, \ \widetilde{h}_2 = h_2 - h_q, \ \ldots, \widetilde{h}_{q-1} = h_{q-1} - h_q.$$

This is where the difference between the properties of the transformation β of the projective space and the properties of the transformation α of a sphere appear—the problem reduces to that of the trajectories on a torus of smaller dimension, under the action of a standard shift, generated by a different set of numbers. Correspondingly, in the transition from sphere to a given projective space, the number of qualitatively different types of trajectories decreases. But, if we consider spheres of different dimensions, then for the action on the corresponding projective space, we have the same possible cases as for the actions on spheres, with the dimension shift by 1.

For example, if the torus T_ξ is a circle, then the torus $T_\xi/\sim$ degenerates into a point, which is fixed for all h.

If the torus T_ξ is two-dimensional, then the torus $T_\xi/\sim$ is a circle, and the action of β on that circle is a rotation. If the number $h_1 - h_2$ is irrational, then the trajectory of each point from T_ξ is dense everywhere. If the number $h_1 - h_2$ is rational, then all the trajectories are periodic with the same period.

Summarizing the above arguments, we obtain a description of dynamics of the map β, on a projective space. This description is very similar to the description of the trajectory on a sphere; it differs in the reduction of the dimension of corresponding tori by 1.

Theorem 5.4.2 *Let A be a nonsingular linear operator in $\mathbb{C}^m$, $\beta : \mathbb{C}P^{m-1} \to \mathbb{C}P^{m-1}$ the transformation of the projective space induced by it, $P_k = \mathbb{C}P^{q(k)-1}$; $k = 1, 2, \ldots, q$, is the projective space consisting of one-dimensional subspaces of the subspace $L^1(k)$, embedded into $\mathbb{C}P^{m-1}$.*

1. *Under the action of the map β, the trajectory of a one-dimensional subspace L spanned by a vector $x \neq 0$, tends to the trajectory of the subspace $[w] \in P_{k(x)}$.*
2. *The subspace P_k stratifies, by the map φ_k, given (5.3.2), into tori $\widetilde{T}(\xi) = \varphi_k^{-1}(\xi)$, being invariant with respect to the action of β.*
3. *Let $J(\xi) = \{j : w(k, j, i) \neq 0\}$, $d(\xi) = |J(\xi)|$ be the number of nonzero coordinates of the point $\xi \in \mathbb{S}_+^{q(k)-1}$ and $m(\xi) = \max\{j : j \in J(\xi)\}$. The dimension of the torus $\widetilde{T}(\xi)$ is equal to $d(\xi) - 1$, and the action of the map β on the torus $\widetilde{T}(\xi)$ coincides with the standard shift of a torus, generated by the element of the torus $\widetilde{\omega}(\xi)$, made up of numbers*

$$\widetilde{\omega}(k, j) = \frac{\lambda(k, j)}{\lambda(k, m(\xi))}, \quad where \ j \in J(\xi), j < m(\xi),$$

and it is described by Theorem 5.2.2.

In particular, the set of limit points of the trajectory of the point $[w]$ is the equivalence class of that point, in the corresponding quotient group $\widetilde{T}(\xi)/H_{\omega(\xi)}$

Corollary 5.4.3 *If all the eigenvalues of A have different moduli, then the trajectory of any one-dimensional subspace has the limit. In particular, this holds if the spectrum of the operator consists of a single point.*

Proof Under the condition of the corollary, all the points from the set P_k are fixed.   □

As it was demonstrated in Example 5.4.1, in the case $m = 2$, the question under consideration is the question on behavior of the trajectory under the action of fractional-linear transformation

$$\beta(z) = \frac{a_{11}z + a_{12}}{a_{21}z + a_{22}}.$$

Let λ_1, λ_2 be the eigenvalues of A and let

$$\lambda = \frac{\lambda_1}{\lambda_2} = Re^{i2\pi h}.$$

As it follows from theorem, the behavior of the trajectory depends on the form of the number λ only.

Let us remark that the behavior of the trajectory under action of fractional-linear transformation was investigated long ago, in a connection with different applications [2, 3]. In particular, in [2] was proposed the following classification of fractional-linear transformation, based on the information on R and h.

The fractional-linear transformation β is said to be of

loxodromic type, if $R \neq 1$, $h \neq 0$ and $h \neq \frac{1}{2}$;

hyperbolic type, if $R \neq 1$, $h = 0$ or $h = \frac{1}{2}$;

elliptic type, if $R = 1$, $h \neq 0$ and $h \neq \frac{1}{2}$;

parabolic type, if $R \neq 1$, $h = 0$ or $h = \frac{1}{2}$.

References

1. Weyl, H. (1984). *Mathematics. Theoretical physics*. Moscow: Nauka.
2. Ford, P. (1936). *Automorphic function*. Moscow: GIFML.
3. Karapetiants, N., Samko, S. (2001). Equations with involutive operators. Berlin: Boston, Basel. Birkhäuser.

Part II
Subspace Trajectory

Chapter 6
Subspaces

The issue on the behavior of the trajectory of a vector subspace, under the action of a linear map, turns out to be more complicated. In particular, the information on the behavior of trajectories of points is not enough for the description of the behavior of subspace trajectories.

The behavior of the trajectory of a subspace V, under the action of the operator A, depends on the position of the given subspace with respect to a series of subspaces occurring in the pyramids (2.2.3) and (2.2.8). This position is given by a number of parameters, and in any concrete problem, one has to clarify which of these parameters it depend on, the property under consideration.

6.1 The Subspace Position with Respect to the Pyramid

We consider the question of how to characterize the position of a subspace with respect to a given decomposition of the space, and with respect to a given pyramid of subspaces.

Let a decomposition of X into a direct sum of subspaces $L(k)$ is given:

$$X = \mathbb{C}^m = \bigoplus_{k=1}^{q} L(k), \qquad (6.1.1)$$

and let V be a given subspace.

To a decomposition (6.1.1), the pyramid of subspaces

$$M(k) = \bigoplus_{i=1}^{k} L(i)$$

is associated. Later, we shall discuss pyramids associated to a given operator A, but in the first propositions this is not used—we any pyramids reconsider.

© Springer International Publishing Switzerland 2014   63
Ć.B. Dolićanin and A.B. Antonevich, *Dynamical Systems Generated by Linear Maps,*
DOI: 10.1007/978-3-319-08228-8_6

Lemma 6.1.1 *Under the given pyramid $M(k)$, an arbitrary subspace V may be represented in the form of a direct sum*

$$V = \bigoplus_k W(k), \tag{6.1.2}$$

where subspace $W(k)$ belongs to the corresponding level of the pyramid:

$$W(k) \subset M(k), \quad W(k) \cap M(k-1) = \{0\}.$$

Proof Subspaces $V_k = V \cap M(k)$ form a nested chain of subspaces. Here, it may happen that for some k one has $V_k = V_{k-1}$, i.e. there are levels with weight 0. According to the Theorem 4.1.6, for this pyramid there is a normal basis, i.e. such a basis that the number of basis vectors lying in the level $S_k = V_k \setminus V_{k-1}$ is equal to the weight of that level. The proposition is satisfied, if $W(k)$ is subspace generated by basis vectors lying in S_k. $\square$

As it can be seen from the proof of the lemma, the dimension d_k of subspaces $W(k)$ is one of the characteristics of the position of the subspace V with respect to the pyramid. Some of these subspaces may be trivial; we have

$$\sum_{k=1}^{m} d_k = d < m.$$

In the text, we repeatedly discuss pairs, consisting of a vector subspace X, and its subspace Y, and in different places we use two forms of complements:

1. *set-theoretic complement $X \setminus Y$*; this set is not a subspace, it is uniquely determined;
2. *algebraic complement*—such a subspace Z in X, so that any vector x from X may, in a unique way, be written as a sum $x = y + z$, $y \in Y$, $x \in Z$. By definition, an algebraic it complement is a subspace; for given X and Y an algebraic complement of Y exists, but is not unique.

Let us also emphasize the difference between the notions of a *sum od subspaces* and of a *direct sum of subspaces*. The sum of arbitrary subspace $V(k)$ is a subspace of the form

$$\sum_{k=1}^{p} V(k) = \{\sum_{k=1}^{p} v_k : v_k \in V(k)\}.$$

In the definition of the direct sum of subspaces $\bigoplus_k V(k)$ the additional requirement is the decomposition of an element in the form $\sum_{k=1}^{p} v_k$ is unique. In particular, for the sum of subspaces, we have

$$\dim \sum_{k=1}^{p} V(k) \le \sum_{k=1}^{p} \dim V(k).$$

The sum of subspaces is a direct sum, if and only if,

$$\dim(\sum_{k=1}^{p} V(k)) = \sum_{k=1}^{p} \dim V(k).$$

In fact, in the proof of Lemma 6.1.1 as a subspace $W(k)$ one takes an algebraic complement of V_{k-1} in V_k, the choice of an algebraic complement is determined by the choice of a normal basis.

There is considerable arbitrariness in the choice of the subspaces $W(k)$, therefore one would like to choose such a subspace with additional properties. We consider two variants of such a choice.

I. We discuss below spaces with a scalar product, where subspaces $L(k)$ are mutually orthogonal. In this situation, one can as $W(k)$ choose the orthogonal complement of V_{k-1} in V_k. However, in the following, it turns out to be essential that the subspace chosen in this was (as orthogonal to V_{k-1}) may not be orthogonal to a bigger subspace $M(k-1)$. Then, the orthogonal projections of the subspace $W(k)$ on the subspaces $L(i)$, for $i < k$ may be nonzero. Non-triviality of these projections reflect the "bad" position of the subspace V with respect to the set of subspaces $L(k)$, and complicates questions being discussed later.

II. Since any vector x is a sum of its projections on subspaces $L(k)$, at first, it may appear that as subspaces $W(k)$ in Lemma 6.1.1, one may take the projections $P(k)(V)$ of V on subspaces $L(k)$, or subspaces

$$\tilde{V}(k) = V \bigcap L(k).$$

However, this is not always true. The posibiliy of decomposing any vector from V is not equivalent to the decomposition of the subspace V itself into a sum of its projections. The difference is, that in the case of a decomposition of a space into a direct sum of projections, there are no relations between different projections $P(k)x$ of vectors $v \in V$. If the subspace V is not decomposable into a direct sum of its projections, then there are relations between various projections which hold for all vectors from V.

The conditions on the decomposability of the subspace V into a direct sum of projections are contained in the next lemma.

Lemma 6.1.2 *I. For any subspace V, there are inclusions*

$$\bigoplus_{k} \{V \bigcap L(k)\} \subseteq V \subseteq \bigoplus_{k} P(k)(V),$$

where both inclusions may be proper. In particular, the inequalities

$$\sum_k \dim(V \cap L(k)) \leq \dim V \leq \sum_k \dim P(k)(V)$$

hold.

    II. The following equalities are equivalent:

i)
$$V = \bigoplus_k P(k)(V),$$

ii)
$$V = \bigoplus_k \{V \cap L(k)\},$$

iii)
$$\dim V = \sum_k \dim P(k)(V),$$

iv)
$$\dim V = \sum_k \dim\{V \cap L(k)\},$$

v)
$$V \cap L(k) = P(k)(V) \ \forall k \ ,$$

vi)
$$P(k)(V) \subset V \ \forall k \ .$$

Proof I The first inclusion is obvious.

    For any vector $x \in V$, there is a decomposition

$$x = \sum_k x(k),$$

where $x(k) = P(k)x \in V(k)$. Therefore,

$$x \in \bigoplus_k V(k),$$

which was required.

Let us give an example showing that equalities may not hold. Consider the decomposition

$$\mathbb{C}^2 = L(1) \bigoplus L(2),$$

where $L(1) = \{(x_1, 0)\}$, $L(2) = \{(0, x_2)\}$, and suppose that $V = \{(x_1, x_1)\}$. Then

$$\{V \bigcap L(1)\} \bigoplus \{V \bigcap L(2)\} = \{0\} \neq V \neq P(1)(V) \bigoplus P(2)(V) = \mathbb{C}^2.$$

Proof II Since

$$V \subseteq \bigoplus_k P(k)(V)$$

and

$$\dim\{\bigoplus_k P(k)(V)\} = \sum_k \dim P(k)(V),$$

the equality (i) may hold if and only if the dimensions of these subspaces are equal, i.e. the condition (ii) holds.

The equivalence of other conditions may be checked by similar elementary considerations. □

As noted before, in questions related to the investigation of asymptotics, graded-linear operators naturally appear. Note that a map $F: X \to X$ is called a graded-linear operator consistent with the pyramid $M(k)$ if in each level $S_k = M_k \setminus M_{k-1}$ the action of F is given by a linear operator T_k and $M_{k-1} \subset \ker T_k$. Note also that graded-linear operators are discontinuous maps and they essentially differ from linear operators. In particular, the image of a linear operator is always a vector subspace, while the image of a graded-linear operator F is a union of vector subspaces—the images of the operators T_k.

Let us show how one can associate to a graded-linear operator F a linear operator reflecting numerous properties of the operator F. This construction shall be used later in the construction of the limit of trajectories of subspaces.

Suppose one has a pyramid of subspaces $M(k)$ and the consistent graded-linear operator F whose action on the level S_k is given by a linear operator T_k. According to Lemma 6.1.1, for subspaces V there is a representation (6.1.2):

$$V = \bigoplus_k W(k),$$

where the subspace $W(k)$ belongs to the corresponding level of the pyramid:

$$W(k) \subset M(k), \quad W(k) \cap M(k-1) = \{0\}.$$

Let $Q(k)$ be the projector on $W(k)$, corresponding to the given subspace decomposition.

Definition 6.1.3 The assembly of operators T_k, induced by the decomposition (6.1.2) of the subspace V, is the linear operator given by the formula

$$T = \sum_{k=1}^{p} T_k Q(k).$$

As it can be seen from the definition, the operator T is formed in such a way that its action on subspaces $W(k)$ agrees with the action of the operator T_k on this subspace. This explains the use of the term "assembly". The relation between the assembly and the initial graded-linear operator F is shown in the following proposition.

Lemma 6.1.4 *Let F be a graded-linear operator and T be the assembly of the corresponding operators T_k. The image $T(V)$ of the operator T is a linear span of the image $F(V)$, so it is the sum of images $T_k(W(k))$.*

Proof The linear span of a set is the smallest vector subspace containing all elements from the given set. In the case under consideration

$$F(V) = \bigcup_k T_k(W(k)),$$

where the sets $T_k(W(k))$ are vector subspaces. Therefore, the linear span of its union is the sum (but not a direct one) of subspaces $T_k(W(k))$.   □

6.2 The Invariant Subspaces

As noted above, the limit of a subspace trajectory under the action of a linear operator is an invariant subspace. Therefore for further study any information concerning the structure of invariant subspaces and the topological structure of the set of invariant subspaces (with the topology induced by the Grassmann manifold) is useful.

Such a description is important in many other problems and the study of invariant subspaces of a linear operator of a finite-dimensional vector space is related to numerous classical problems of linear algebra. In particular, well known monograph [1] is dedicated to this theme. We give some facts are necessary in what follows later. Some of these facts are known and are contained in, for example, just mentioned monograph [1]. Other facts had to be obtained independently or reformulated in the form more convenient for our further investigation.

In the following, we assume that the previous assumptions hold and use already given notation and objects. We remark that we discuss an invertible operator A in m-dimensional complex space X. To this operator, decompositions (2.2.2), the pyramids (2.2.3) and (2.2.8) are associated. Let us clarify the arrangement of invariant subspaces with respect to these decompositions.

Lemma 6.2.1 *A subspace V is invariant with respect to the operator A, if and only if, it may be represented as a direct sum of subspaces*

$$V = \bigoplus_{k=1}^{q} \bigoplus_{j=1}^{q(k)} V(k, j), \qquad (6.2.3)$$

where $V(k, j)$ is an invariant subspace in $L(k, j)$.

One of possible proofs of this lemma may be found in [1]. This proof uses a number of preceding facts, and it is difficult to reproduce it here, completely. For completeness and independence of the presentation, we give another proof.

Proof It is obvious that the condition is sufficient.

Necessity. Let V be an arbitrary invariant subspace. Since the subspace $L(k, j)$ is invariant, the intersection $V(k, j) = V \cap L(k, j)$ is also an invariant subspace. We show that the equality (6.2.3) holds. According to Lemma 6.1.2, we have an inclusion

$$\bigoplus_{k,j} V(k, j) \subset V,$$

where the equality (6.2.3) is equivalent to the fact that for any k and l

$$P(k, j)(V) \subset V. \qquad (6.2.4)$$

The last inclusion may be checked by the following known construction.

Let f be an analytic function of complex variable, defined in a neighborhood of every eigenvalue of the matrix A. As it is known [2], the value $f(A)$ of the function f at the matrix A is given by

$$f(A) = \frac{1}{2\pi i} \oint_{\Gamma} (\lambda I - A)^{-1} f(\lambda) d\lambda,$$

where Γ is a closed contour, containing within itself all eigenvalues of the matrix A and such that on the region, bounded by that contour, the function f is analytical. On the other hand, any operator of the form $f(A)$ may be represented as a polynomial of the matrix A, i.e., there is a polynomial $Q(\lambda)$, where $f(A) = Q(A)$. This polynomial is given as follows. Let the minimal polynomial of the matrix A has the form

$$\psi(\lambda) = (\lambda - \lambda_1)^{m_1} (\lambda - \lambda_2)^{m_2} \ldots (\lambda - \lambda_s)^{m_s}.$$

Note that the number m_k coincides with the maximal among the dimensions of the Jordan cells corresponding to the eigenvalue λ_k. Then, the polynomial $Q(\lambda)$ is obtained from the conditions

$$Q(\lambda_k) = f(\lambda_k), \ Q'(\lambda_k) = f'(\lambda_k), \ldots, Q^{(m_k-1)}(\lambda_k)$$
$$= f^{(m_k-1)}(\lambda_k), k = 1, 2, \ldots, s.$$

and it is the Lagrange–Sylvester interpolation polynomial for the function f [2].

Consider the function $f_{k,j}(\lambda)$, whose values are 1 in a neighborhood of the eigenvalue $\lambda(k, j)$ and $f_{kj}(\lambda) = 0$ in a neighborhood of other eigenvalues. Then the operator $f_{kj}(A)$ is a projector onto the subspace $L(k, j)$, i.e., $f_{kj}(A) = P(k, j)$. According to previous results, the projector $P(k, j)$ may be represented as a polynomial of the operator A: $P(k, j) = Q_{kj}(A)$, where $Q_{kj}(\lambda)$ is some polynomial.

From this, claiming of the lemma follows. Indeed, by the invariance of the subspace V, all vectors $A^n x$ are in V, therefore the vector $P(k, j)x$, being their linear combination, also belongs to V. So, (6.2.4) holds.   □

For example, if the operator A has m different eigenvalues, then all subspaces $L(k)$ are one-dimensional. Therefore, non-zero invariant subspaces of $L(k)$ coincide with $L(k)$ and from Lemma 6.2.1 it follows that there are only finitely many invariant subspaces. Namely, these are exactly those subspaces which are representable in the form of a direct sum of some collection of one-dimensional subspaces $L(k)$.

We remark, that although the decomposition

$$L(k, j) = \bigoplus_{i=1}^{q(k,j)} L(k, j, i)$$

holds, the invariant subspace $V(k, j)$ may not be a direct sum of its projections $P(k, j, i)(V(k, j))$ onto the subspaces $L(k, j, i)$, i.e. in general

$$V(k, j) \neq \bigoplus_{i=1}^{q(k,j)} P(k, j, i)V(k, j, i).$$

This shall be shown in examples. Therefore, the lemma may not be strengthen in the general case—it is impossible to claim that an invariant subspace is a direct sum of invariant subspaces belonging to $L(k, j, i)$, and one cannot reduce the problem of the description of invariant subspaces to the problem of the description of invariant subspaces in $L(k, j, i)$. In particular, this is the reason that exactly the case in which one has several Jordan cells corresponding to the same eigenvalue, is the most complex in the description of invariant subspaces and in the study of trajectories of an arbitrary subspace.

6.3 The Invariant Subspaces of an Operator with One-Point Spectrum

In the general case the action of the operator A on each subspace $L(k, j)$, is given by the set of the Jordan cells with the same eigenvalue. In other words, the spectrum of the operator $A(k, j)$, introduced above, consists of one point $\lambda(k, j)$. Therefore, according to Lemma 6.2.1, the problem of the description of invariant subspaces of the operator A is reduced to the description of subspaces invariant with respect to the action of an operator, whose spectrum consists of one point.

We consider such a case in detail and discuss how extensive may be the set of invariant subspaces.

Let us consider an operator B in a finite-dimensional space X, such that the spectrum of B has only one point λ. Operator B has the form $B = \lambda I + N$, where N is a nilpotent operator. Under the previously introduced numeration, first two indices k and j may only assume value 1, so for the simplicity of the notation, in the following, we use the numeration of basis vectors by two last indices i and l.

In this way, we consider a basis $e(i, l)$, indexed by two indices: $1 \leq i \leq q$, $1 \leq l \leq q(i)$ where i is the index of the Jordan cell, $q(i)$ is the dimension of that cell. Here the index l shows what is the order of the vector $e(i, l)$, as an adjoint vector. The quantity l may assume values from 1 to $\widetilde{q}$, where $\widetilde{q} = \max_i q(i)$.

The operator N acts by formulas $Ne(i, l) = e(i, l - 1)$, if $l > 1$; $Ne(i, 1) = 0$.

Let $M(l)$ be a subspace spanned by the vectors $e(i, l)$ with given l; if $l > 1$, the operator N injectively maps subspace $M(l)$ into $M(l - 1)$; $N(M(1)) = \{0\}$.

Subspaces

$$M^l = \bigoplus_{k=1}^{l} M(k), \ l = 1, 2, \ldots, \widetilde{q};$$

form a pyramid, induced by the operator B. Note that the pyramid of subspaces M^l is canonical—it is uniquely determined by the operator B.

For an arbitrary subspace V, intersections

$$V_l = V \bigcap M^l$$

form a linearly ordered collection of subspaces in V. But, some of these subspaces may be identical—this collection of subspaces may turn out to be a degenerated pyramid, in particular, starting from some $p = p(V)$, we have $V_p = V$.

Lemma 6.3.1 *Subspace V is invariant if and only if*
i) $V_l \neq V_{l-1}$ *for* $l \leq p$,
ii) $N(V_l) \subset V_{l-1}$.

Proof The condition, that $V_p \neq V_{p-1}$ means that in the space V there is an adjoint vector z of order $p - 1$. Then, from the invariance of the subspace V, this subspace contains adjoint vectors of all orders less than $p - 1$. Indeed, we have $N^k z \in V$, and this vector is an adjoint vector of order $p - 1 - k$. The equality $V_l = V_{l-1}$ is impossible since it is equivalent to the fact that in V there are no adjoint vectors of order $l - 1$.   □

Let us clarify the meaning of the condition (ii) in Lemma 6.3.1. The subspace V_{l-1} may be given by a certain set of linear relations on coordinates $x(i, j)$ of vectors from that subspace:

$$V_{l-1} = \{x \in M^{l-1} : \sum_{ij} a_{ij}^k x(i, j) = 0, \ k = 1, \ldots, \varrho; \ \}.$$

Then the condition (ii) means that for the coordinates of vectors from the subspace V_l a linear relations with the same coefficients must hold for the coordinates of vectors from the subspace V_l with the change of the index j of the coordinate with the index $j + 1$:

$$\sum_{ij} a_{ij}^k x(i, j + 1) = 0, \ k = 1, \ldots, \varrho.$$

Naturally, among the coordinates of vectors from the subspace V_l some other relations may also hold.

For an arbitrary subspace V, the intersections

$$V_l = V \bigcap M^l$$

form linearly ordered set of subspaces of V, but the inclusion $N(V_l) \subset V_{l-1}$ need not hold. Consider the numbers

$$\delta_k = \dim\{N(V_l)/[N(V_l) \bigcap V_{l-1}]\} = \dim N(V_l) - \dim[N(V_l) \bigcap V_{l-1}].$$

Lemma 6.3.1 claims that the subspace V is invariant, if and only if, all $\delta_k = 0$ and the set of numbers δ_k characterizes qualitative difference of the subspace V from an invariant one.

Lemma 6.3.1 may be used for the construction and description of all invariant subspaces. We construct step-by-step subspaces forming a pyramid $V_1 \subset V_2 \subset \ldots \subset V_p$ and satisfying conditions

$$V_l \subset M^l, \ N(V_l) \subset V_{l-1}. \tag{6.3.5}$$

As V_1, we may take any subspace of $M(1)$ (in $M(1)$ all subspaces are invariant). If $q = \dim M(1) = 1$ (i.e. there is only one Jordan cell), there is only one such a subspace $V_1 = M(1)$. If $q > 1$, there are infinitely many such subspaces.

Let $\tilde{V}_1 = N^{-1}(V_1) \subset M^2$ be the full inverse image of V_1. This subspace is invariant and $V_1 \subset \tilde{V}_1$, since $N(V_1) = 0$. An arbitrary subspace V_2, such that

$V_2 \neq V_1$ and $V_1 \subset V_2 \subset \widetilde{V}_1$ is invariant. Note that if $q = 1$, then there is only one such a subspace.

We proceed analogously. Let $\widetilde{V}_2 = N^{-1}(V_2) \subset M^3$ be the full inverse image of V_2. An arbitrary subspace V_3, such that $V_3 \neq V_2$ and $V_2 \subset V_3 \subset \widetilde{V}_2$ is invariant.

Continuing with this process, we construct subspaces V_k such that

$$V_k \neq V_{k-1}, \quad V_{k-1} \subset V_k \subset \widetilde{V}_{k-1}. \tag{6.3.6}$$

Note that if $\dim \widetilde{V}_{k-1} = \dim V_{k-1}$, the required subspace V_k does not exist and the process terminates. If $\dim \widetilde{V}_{k-1} - \dim V_k = 1$, then there is only one subspace $V_k = \widetilde{V}_{k-1}$. In all other cases, there are many possibilities for the choice of V_k. As a result of these constructions, we obtain all invariant subspaces, and in the general case, the operator B has very extensive family of invariant subspaces.

The process of construction of invariant subspaces may be described in other terms, using the decomposition from Lemma 6.1.1.

As before, we consider the pyramid of subspaces M^l, induced by the operator B. According to Lemma 6.1.1, for any subspace V, there is a decomposition into a direct sum of subspaces

$$V = \bigoplus_1^p W(l), \tag{6.3.7}$$

where

$$W(l) \subset M^l, \quad W(l) \bigcap M^{l-1} = \{0\}.$$

In this case,

$$V_l = \bigoplus_{k=1}^{l} W(k) = V_{l-1} \bigoplus W(l).$$

Lemma 6.3.2 *Suppose that the spectrum of the operator $B = \lambda I + N$, in a finite-dimensional space L, consists of one point λ.*

The subspace V is invariant, if and only if, under the decomposition (6.3.7), the following conditions are satisfied

$$W(l) \neq 0, \quad N(W(l)) \subset \bigoplus_{k=1}^{l-1} W(k).$$

According to this Lemma, for the construction of the next subspace V_l, one has to find subspaces $W(l)$, satisfying the stated conditions.

From the previous Lemmas, we obtain the following description of an arbitrary invariant subspace.

Theorem 6.3.3 *A subspace V is invariant with respect to a linear operator A in a finite-dimensional space, if and only if, this subspace may be represented in the form of a direct sum*

$$V = \bigoplus_{k,j} \bigoplus_{l} W(k, j; l),$$

where subspaces $W(k, j, l)$, *for fixed k and j, satisfy conditions:*

$$W(k, j, 1) \neq 0, \quad N(k, j)(W(k, j, l)) \subset \bigoplus_{i=1}^{l-1} W(k, j, i).$$

6.4 The Set of Invariant Subspaces for a Model Example

As an example, we consider the linear operator in the space $X = \mathbb{C}^4$, given by the matrix

$$A = \begin{pmatrix} 1 & 1 & 0 & 0 \\ 0 & 1 & 0 & 0 \\ 0 & 0 & 1 & 1 \\ 0 & 0 & 0 & 1 \end{pmatrix}, \tag{6.4.8}$$

having two identical Jordan cells. Here, the vectors $(1, 0, 0, 0)$ and $(0, 0, 1, 0)$ are eigenvectors, and the vectors $(0, 1, 0, 0)$ and $(0, 0, 0, 1)$ are adjoint vectors.

The pyramid corresponding to this operator is $M^1 \subset M^2 = X$, where the subspace M^1 is spanned by eigenvectors $(1, 0, 0, 0)$ and $(0, 0, 1, 0)$.

We find all two-dimensional subspaces in $\mathbb{C}^4$ invariant, with respect to the given operator, and we describe the set $F_2(A)$ of such subspaces.

For the construction of an arbitrary two-dimensional invariant subspace we apply Lemma 6.3.2. We look for V in the form

$$V = W(1) \bigoplus W(2), \tag{6.4.9}$$

where

$$W(1) \subset M^1, \quad W(2) \bigcap M^1 = \{0\}, \quad N(W(2)) \subset W(1).$$

One two-dimensional invariant subspace, which we denote by V_0, is constructed easily—this is the case when dim $W(1) = 2$, dim $W(2) = 0$. Then $V_0 = W(1) = M^1$.

For all other invariant subspaces dim $W(1) = 1$, dim $W(2) = 1$, so the case dim $W(1) = 0$, dim $W(2) = 2$ is impossible. Let us construct all these subspaces.

Lemma 6.4.1 *A subspace* $V \neq V_0$ *is invariant, if and only if, it is generated by vectors* $v = (z_1, z_2, z_3, z_4)$ *and* $v_1 = (z_2, 0, z_4, 0)$, *where* $|z_2| + |z_4| \neq 0$.

Proof As $W(2)$, we take a one-dimensional subspace spanned by the non-zero vector $v = (z_1, z_2, z_3, z_4)$. From the conditions $W(2) \cap M^1 = \{0\}$, we obtain that $|z_2| + |z_4| \neq 0$. From the condition $N(W(2)) \subset W(1)$, we get that the subspace $W(1)$ is spanned by the vector $v_1 = Nv = (z_2, 0, z_4, 0)$.   □

The correspondence between vectors v and invariant subspaces $V \neq V_0$ described in the Lemma, is not bijective, since different vectors v may correspond the same V. Therefore, such vectors should be considered equivalent, and the set of invariant subspaces may be described as a quotient space by the corresponding equivalence relation. Under such description, some objects from the theory of vector bundles are used. We recall the necessary facts and definitions.

6.5   Vector Bundles

One says that on a topological space Y, is given the structure of *a bundle* with base X, if one has a continuous surjective map $p : Y \rightarrow X$ (which is called *the projection*). The space X is called *the base space*, the space Y is called *the total space*, and the set $p^{-1}(x)$ is called the *fiber* of the bundle over the point x.

The bundle $p : Y \rightarrow X$ is called *a locally trivial fibre bundle* with the typical fibre F if each point in X has a neighborhood W, where the inverse image $p^{-1}(W)$ is homeomorphic to the product $W \times F$, and the homeomorphism commutes with projections.

An example of a fiber bundle is a direct product $Y = X \times F$ (product-bundle). If the space X is a ball in a finite-dimensional space, then any fiber bundle is trivial—isomorphic to a product-bundle. But, for more complex space X, there may be non-trivial fiber bundles Y. Such spaces Y have more complex topological structure than the product $X \times F$. The fact that the space Y has the structure of a bundle helps in studying its topological properties. In particular, there is a connection between homotopy groups of space X, Y and F given in the form of the so-called exact sequence [3, 4].

One of the methods of description of bundles is by specifying their gluing functions.

Let W_i be those open subsets of the base space X for which the inverse image $p^{-1}(W_i)$ is homeomorphic to the product $W_i \times F$ and let $\varphi_i : p^{-1}(W_i) \rightarrow W_i \times F$ being the corresponding homeomorphisms.

Let $W_{ij} = W_i \cap W_j$. If the set W_{ij} is non-empty, then there is a map $\varphi_{ij} = \varphi_i \circ \varphi_j^{-1}$ of the set $W_{ij} \times F$ into itself. This map is a homeomorphism and, since it commutes with projections, it is of the form

$$\varphi_{ij}(x, \xi) = (x, \psi_{ij}^x(\xi)); \ x \in W_{ij}, \xi \in F,$$

where for fixed x, the map $\psi_{ij}^x(\cdot)$ is a homeomorphism of the fiber F. In this way, on the intersections $W_i \cap W_j$, one has maps ψ_{ij}^x into the group of homeomorphisms of the fiber F. These maps are called *the gluing functions*. If the family of sets W_i is a covering of the space X, then the gluing functions determine the bundle up to an isomorphism.

If for each x all gluing functions ψ_{ij}^x belong to a certain subgroup G of the group of all homeomorphisms of the space F, then G is called *the structure group of the bundle*. Usually, a subgroup G of the group of all homeomorphisms of the space F occurs in these situations, when on the typical fiber F there is an additional structure; then for group G one takes the subgroup containing homeomorphisms of the fiber preserving the given structure.

Particularly important is the case when the typical fiber is a vector space and the group G is the group of linear maps. The bundle $p: Y \to X$ is called *an n-dimensional vector bundle* if the typical fiber F is an n-dimensional vector space and the structure group is the group of invertible linear maps of the fiber F.

One of the first results in vector bundle theory is the description of all one-dimensional complex vector bundles over two-dimensional sphere S^2.

Recall that the Riemann sphere (extended complex plane) is the set $\mathbf{C}^+ = \mathbb{C} \bigcup \{\infty\}$ with the corresponding topology. As a topological space, the Riemann sphere is homemorphic to the ordinary two-dimensional sphere S^2, therefore we consider bundles over the Riemann sphere.

Let

$$D_1 = \{z \in \mathbb{C} : |z| \le 1\}$$

be the unit disc in the complex plane. The set

$$D_2 = \{z \in \mathbf{C}^+ : |z| \ge 1\}$$

is also a disc. The Riemann sphere is obtained from D_1 and D_2 by gluing the corresponding points on the unit circle.

Let Y be an arbitrary one-dimensional vector bundle over the Riemann sphere. Over the disc D_1, it is trivial and it is isomorphic to the product

$$D_1 \times \mathbb{C} = \{(z, \xi_1) : z \in D_1, \xi_1 \in \mathbb{C}\}.$$

Analogously, over the disc D_2 the bundle is trivial and it is isomorphic to the product

$$D_2 \times \mathbb{C} = \{(z, \xi_2) : z \in D_2, \xi_2 \in \mathbb{C}\}.$$

Therefore, Y may be obtained by gluing these two product-bundles. Such gluing may be achieved as follows.

Let k be an arbitrary integer. We identify the point $(z_1, \xi_1) \in D_1 \times \mathbb{C}$ with the point $(z_2, \xi_2) \in D_2 \times \mathbb{C}$ if $|z_1| = |z_2| = 1$, $z_1 = z_2 = z$ and $\xi_1 = z^k \xi_2$. The space E_k^2

obtained by this gluing is a one-dimensional complex vector bundle over the sphere. The bundle E_1^1 is called the Hopf bundle, and it is usually denoted by H.

Proposition 6.5.1 *Each one-dimensional complex vector bundle over the sphere S^2 is isomorphic to one of the bundles E_k^1, and for different k these bundles are non-isomorphic.*

The existence of non-trivial bundles over S^2 was established by H. Hopf in the proof of the fact that the homotopy group $\pi_3(S^2)$ is non-trivial. This result of Hopf was one of the mathematical sensations of the thirties of twentieth century (see [5]).

The map $f : S^3 \to S^2$ constructed by Hopf and giving a generator of the group $\pi_3(S^2)$, may be described as follows.

It is convenient to consider the space $\mathbb{R}^4$ as $\mathbb{C}^2 = \{(z, w) : z, w \in \mathbb{C}\}$. Under this identification, the use of the action by complex numbers allows more compact expressions.

The three-dimensional sphere is realized as

$$S^3 = \{(z, w) : |z|^2 + |w|^2 = 1\} \subset \mathbb{C}^2.$$

On S^3, one gives an equivalence relation. The point (z, w), is considered equivalent to any point of the form $(\lambda z, \lambda w)$, where $|\lambda| = 1$. Any equivalence class has the same structure as the circe. The quotient space by this equivalence relation is a two-dimensional sphere. Indeed, the equivalence class $[(z, w)]$ is uniquely determined by the quotient $u = z/w$, where this quotient may assume value ∞, i.e. this is a point from the extended complex plane—the Riemann sphere.

In this way, one has the map $f : S^3 \to S^2$, given by the formula

$$f(z, w) = [(z, w)],$$

that is exactly the Hopf map.

This construction gives a locally trivial bundle of the sphere S^3 over S^2, where the typical fiber is the circle S^1. One-dimensional complex vector bundle H, given above, is essentially related to the Hopf map.

We present another method of the construction of this bundle. Consider the product

$$S^2 \times \mathbb{C}^2 = \{([(z, w)], z_1, z_2) : [(z, w)] \in S^2, z_1, z_2 \in \mathbb{C}\}$$

and its subset

$$H = \{([(z, w)], \lambda z, \lambda w) : \lambda \in \mathbb{C}\}.$$

Here, one has the natural projection $p : H \to S^2$, as the projection on the first coordinate: $p(u, z_1, z_2) = u$. The inverse image of each point under this map is a one-dimensional complex subspace, i.e. the map p specifies on H the structure of an one-dimensional complex vector bundle. Simple calculations show this bundle is isomorphic to the Hopf bundle.

The Hopf bundle is of great importance in the number of problems, in particular in K-theory, this bundle induces a generator in the ring $K(S^2)$. We show this bundle occurs in the study of such a simple question as the description of the set of invariant subspaces for the model operator (6.4.8).

6.6 The Structure of Invariant Subspaces Set for a Model Example

Theorem 6.6.1 *The set $\widetilde{F}_2(A)$, consisting of two-dimensional invariant subspaces for the model operator (6.4.8), different from the subspace $V_0 = M^1$ is a two-dimensional complex manifold and it has the structure of a one-dimensional vector bundle on the sphere S^2, isomorphic to the Hopf bundle.*

Proof According to Lemma 6.4.1 every invariant subspace V is spanned by the basis vectors $v = (z_1, z_2, z_3, z_4)$ and $v_1 = (z_2, 0, z_4, 0)$, where $|z_2| + |z_4| \neq 0$.

The vector $v_1 = (z_2, 0, z_4, 0)$, determines a one-dimensional subspace $W(1)$, it is determined up to the multiplier and the parameter

$$z = \frac{z_2}{z_4} \in \mathbb{C}^+$$

uniquely determines the subspace $W(1)$.

With the given z, we shall choose a basis vector v_2 in V in a special way.

If $z \neq \infty$, there is a unique basis vector v_2 having the form

$$v_2 = (\xi_1, z, 0, 1),$$

where $\xi_1 = z_1 - z_3 z$. Thus, each invariant subspace, corresponding to $|z| \leq 1$, is given by a pair of complex numbers

$$(z, \xi_1) \in D_1 \times \mathbb{C}.$$

For $z \neq 0$, there is a unique basis vector v_2, having the form

$$v_2' = (0, 1, -\xi_2, 1/z),$$

where $\xi_2 = z_1/z - z_3$. Thus, each invariant subspace corresponding to $|z| \geq 1$, is given by a pair of complex numbers

$$(z, \xi_2) \in D_2 \times \mathbb{C}.$$

This discussion shows that on the set of invariant subspaces, one has the natural structure of two-dimensional complex manifold with the structure of locally trivial bundle over the Riemann sphere, with the typical fiber $\mathbb{C}$.

Let us find the corresponding gluing function. For this, one has for $|z| = 1$ to find the relation between ξ_1 and ξ_2. But, this relation is obvious:

$$\xi_1 = z_1 - z_3 z = \xi_2 z.$$

So, the gluing function, corresponding to the bundle under consideration is $\varphi(z) = z$, and this bundle is isomorphic to the Hopf bundle. □

References

1. Gohberg, I., Lancaster, P., Rodman, L. (1986). *Invariant subspace of matrices with applications*. New York, Chichester, Toronto, Singapore: Willey.
2. Gantmakher, F. R. (1988). *Theory of matrices*. Moscow: Nauka.
3. Husemoller, D. (1966). *Fibre bundles*. New York, St.Louis, San Francisco, Toronto, London, Sydney: McGraaw-Hill Book Company.
4. Mishchenko, A. S. (1984). *Vectors bundles and its applications*. Moscow: Nauka.
5. Fuks, D. B., Fomenko, A. T., & Gutenmakher, V. L. (1969). *Homotopic topology*. Moscow: MGU.

Chapter 7
The Action of a Linear Map on the Grassmann Manifold

The set of subspaces of a given vector space forms the Grassmann manifold, a linear non-singular operator A induces a homeomorphism φ_A of the Grassmann manifold. Let us find the explicit form or this map.

7.1 The Grassmann Manifold

Let us first present some necessary information on the Grassmann manifold.

Recall that *a differentiable m-dimensional real manifold* is the Hausdorff topological space M, with the following structure:

(1) M may be represented as a union of finitely or countably many open subsets U_k, such that each U_k is homeomorphic to a certain domain V_k of the Euclidean space $\mathbb{R}^m$;

(2) For each U_k there is a homemorphism $\varphi_k : U_k \mapsto V_k \subset \mathbb{R}^m$. The set U_k is called *chart*, coordinates of the vector $\varphi_k(\xi) \in \mathbb{R}^m$ are called *local coordinates* of the point $\xi \in M$. The collection of pairs $\{(U_k, \varphi_k)\}$ is called *an atlas* on the manifold.

(3) If the intersection $U_{kj} = U_k \cap U_j$ is non-empty, there are two systems of local coordinates given by maps φ_k and φ_j respectively; it is required that any of these local coordinates is expressed by other in a differentiable form. This means that the map $\varphi_{kj} : \varphi_k(U_{kj}) \to \varphi_j(U_{kj})$, given by the formula $\varphi_{kj} = \varphi_j \circ \varphi_k^{-1}$, is differentiable and its Jacobian is different from zero at each point from $\varphi_k(U_{kj})$.

Functions φ_{kj} are called *transition functions* (from coordinates to coordinates). If all transition function φ_{kj} are continuously differentiable l times, then the quantity l is called the class of smoothness of the manifold M (it may happen that $l = \infty$).

The atlas $\{(U_k, \varphi_k)\}$ is called *oriented*, if the local coordinates chosen in such a way that Jacobians of all transition functions are positive. If such atlas exists, the manifold is called *an orientable manifold*. There are both orientable and non-orientable manifolds.

© Springer International Publishing Switzerland 2014

81

Ć.B. Dolićanin and A.B. Antonevich, *Dynamical Systems Generated by Linear Maps*, DOI: 10.1007/978-3-319-08228-8_7

Let one has two manifolds: $M = \cup U_k$ with local coordinates x_k^α, and $N = \cup W_i$ with local coordinates y_i^β. We consider a continuous map $f : M \to N$. For a given point $\xi_0 \in M$, the point $f(\xi_0) \in N$ lies in one of the charts W_i and by continuity, for all ξ from a small enough neighborhood U of the point ξ_0, the images $f(\xi)$ belong to the neighborhood W_i, and one may assume that the neighborhood is in one of the charts U_k. Then, in the neighborhood U, the map f may be given in local coordinates, i.e. the expression $f_{kj} = \psi_i \circ f \circ \varphi_k^{-1}$ determines a map from $\varphi_k(U) \subset \mathbb{R}^m$ into $\psi_i(W_i) \subset \mathbb{R}^m$. The map f is called *smooth of class l*, if all the maps f_{kj} are smooth of class l.

The definition of a complex manifold is analogous. The difference appear in the definition of the notion of a differentiable manifold. The space $\mathbb{C}^m$ may be regarded as a real space $\mathbb{R}^{2m}$, and the complex manifold may be considered as a real manifold (with twice dimension). The differentiable map into $\mathbb{C}^m$ is differentiable as a map from $\mathbb{R}^{2m}$ into $\mathbb{R}^{2m}$, but the reverse is not true. Therefore, the class of $2m$-dimensional real differentiable manifolds is broader than the class of m-dimensional complex differentiable manifolds.

Let us show how one gives the manifold structure on the set $G(m, d)$ of all d-dimensional vector subspaces of the space $\mathbb{C}^m$.

Let $G(m, d)$ be the set consisting of all d-dimensional vector subspaces of the space $\mathbb{C}^m$ and

$$G(m) = \bigcup_d G(m, d)$$

—the set of all subspaces in $\mathbb{C}^m$. On the set $G(m)$, there is a topology corresponding to the intuitive notion of closeness of subspaces and turning it into the Hausdorff topological space. For example, two d-dimensional subspaces may naturally be considered close, if they contain normed bases e_k and e'_k respectively, such that $\|e_k - e'_k\| < \varepsilon$ for small ε. This leads to the definition of a neighborhood of the given subspace, i.e. to the definition of a topology.

A topology on the set $G(m)$ may be given using a metric. Any linear operator A acting on $\mathbb{C}^m$ is given by a matrix of format $m \times m$. We denote by $\mathbb{C}^{m \times m}$ the space of all $m \times m$ matrices, and we identify this space with the space of all linear operators on $\mathbb{C}^m$. A norm on the space $\mathbb{C}^m$ induces the norm

$$\|A\| = \max_{\|x\|=1} \|Ax\|$$

on $\mathbb{C}^{m \times m}$, induces the metric $\varrho(A, B) = \|A - B\|$ and induces the corresponding topology on the space $\mathbb{C}^{m \times m}$.

Since all norms on a finite-dimensional space are equivalent, the topology on the space $\mathbb{C}^{m \times m}$ is uniquely determined—it does not depend on the choice of a norm.

We assume, that a scalar product on the space $\mathbb{C}^m$ is given (and we use the Euclidean norm $\|x\| = (x, x)^{1/2}$, induced by the scalar product). Then for any subspace V, it is uniquely determined operator P_V of orthogonal projection on that subspace; thus the map

$$P : V \mapsto P(V) = P_V$$

determines an embedding of the set $G(m)$ into the space $\mathbb{C}^{m \times m}$, and on $G(m)$ a metric is defined, induced from $\mathbb{C}^{m \times m}$:

$$\varrho(V_1, V_2) = \|P_{V_1} - P_{V_2}\| = \max_{\|x\|=1} \|P_{V_1}x - P_{V_2}x\|.$$

In such a way, the set $G(m)$ is a metric space and, in particular, it is the Hausdorff topological space.

An operator is an operator of orthogonal projection onto the non-trivial subspace, if and only if the conditions

$$P^2 = P, \|P\| = 1$$

hold. The self-adjointness of the operator P follows from these conditions. Therefore, identifying a subspace with the projector onto that subspace, we obtain that

$$G(m) = \{P \in \mathbb{C}^{m \times m} : P^2 = P, \ \|P\| = 1\}.$$

In particular, it follows from the obtained representation that the subset $G(m)$ is bounded and closed in $\mathbb{C}^{m \times m}$. Consequently, $G(m)$ is a compact space.

The number

$$\Theta(V_1, V_2) = \max\{\nu(V_1, V_2), \nu(V_2, V_1)\},$$

where

$$\nu(V_1, V_2) = \max_{x \in V_1, \|x\|=1} \quad \min_{y \in V_2,} \|x - y\|,$$

is called *a gap between subspaces*. A gap, in the general case does not give a metric on the set of subspaces, since the triangle inequality need not hold. We remark that for a gap, the inequality $\Theta(L, M) \leq 1$, is obvious.

Lemma 7.1.1 *If the norm on $\mathbb{C}^m$ is induced by a scalar product, then*

$$\Theta(L, M) = \varrho(L, M)$$

and, in particular, the inequality $\varrho(L, M) \leq 1$ holds.

Proof Let us first assume that the projection of a vector onto the subspace M is the closest element in that subspace. Therefore,

$$\min_{y \in M,} \|x - y\| = \|x - P(M)x\|$$

and

$$\nu(L, M) = \max_{x \in L, \|x\|=1} \min_{y \in V_2,} \|x - y\| = \max_{x \in L, \|x\|=1} \|x - P(M)x\|$$
$$= \|[I - P(M)]P(L)\|.$$

Thus,

$$\Theta(L, M) = \max\{\|[I - P(M)]P(L)\|, \ \|[I - P(L)]P(M)\|\}.$$

In addition to that, operators $I - P(M)$ and $P(L)$ are self adjoint, and

$$\|[I - P(M)]P(L)\| = \|\{[I - P(M)]P(L)\}^*\| = \|P(L)[I - P(M)].$$

If $x \in L$, then $[I - P_M]P(L)x = P(L)x - P(M)x$, where

$$\|[I - P(M)]P(L)\| \leq \|P(L) - P(M)\|.$$

Analogously,

$$\|[I - P(L)]P(M)\| \leq \|P(L) - P(M)\|.$$

In this way,

$$\Theta(L, M) \leq \|P_L - P_M\|.$$

On the other hand, for each x, there is a decomposition

$$P(L)x - P(M)x = P(L)[I - P(M)]x - [I - P(L)]P(M)x,$$

where the summands are orthogonal. Therefore,

$$\|P_L x - P_M x\|^2 = \|[I - P(L)]P(M)x\|^2 + \|P(L)[I - P(M)]x\|^2.$$

From the last inequality, we get

$$\|P(L)x - P(M)x\|^2 \leq \max\{\|P(L)[I - P(M)]\|, \ \|[I - P(L)]P(M)\|\}\|x\|^2,$$

being required to.

In singular cases, when $P(M)x = 0$ or $[I - P(M)]x = 0$, the inequality obviously holds. $\qquad\square$

These calculations show that, in the calculation of the norm

$$\|P(L) - P(M)\| = \max_{\|x\|=1} \|P(L)x - P(M)x\|,$$

one may restrict oneself to the vectors lying in the union of subspaces

$$\|P(L) - P(M)\| = \max_{\|x\|=1, x \in L \cup M} \|P(L)x - P(M)x\|.$$

Lemma 7.1.2 *Equality* $\Theta(L, M) = 1$ *holds if and only if one of the subspaces contains a vector x orthogonal to the other subspace.*

Proof Suppose that $\Theta(L, M) = 1$, for example, $\nu(L, M) = 1$. This is equivalent to the fact that there is a vector, $x \in L$, $\|x\| = 1$, such that $\|x - P(M)x\| = 1$. Since $1 = \|x\|^2 = \|x - P(M)x\|^2 + \|p(M)\|^2$, the previous equality is equivalent to $P(M)x = 0$, i.e. the vector x from L is orthogonal to the subspace M. $\qquad\square$

Corollary 7.1.3 *(The Szőkefalvi-Nagy Theorem). If* $\Theta(L, M) < 1$, *then* $\dim L = \dim M$.

The projector $P(M)$ bijectively maps subspace L onto the subspace M if and only if $\Theta(L, M) < 1$.

Proof If $\dim L > \dim M$, then $L \cap M^{\perp} \neq \{0\}$, i.e. in L there is a non-zero vector orthogonal to the subspace M, where $\Theta(L, M) = 1$. Therefore, for $\Theta(L, M) < 1$, we have $\dim L = \dim M$. In addition to that,

$$\{x \in L : P(M)x = 0\} = \{0\}.$$

This means that the projector $P(M)$ acts on L injectively, therefore the image L is the whole subspace M. $\qquad\square$

It follows from this that the set $G(m, d)$ of subspaces of a given dimension d is closed (therefore compact) subset in $G(m)$.

The structure of a smooth manifold on $G(m, d)$ is given as follows. For a given subset V_0, there is a decomposition of the space $\mathbb{C}^m$ into a direct sum $\mathbb{C}^m = V_0 \oplus V_0^{\perp}$, i.e. each vector x is uniquely represented in the form of the sum $x = x_1 + x_2$, $x_1 \in V_0$, $x_2 \in V_0^{\perp}$ (where $x_1 = P_{V_0}x$). This means that the space $\mathbb{C}^m$ may be identified with the set of pairs $(x_1, x_2) : x_1 \in V_0$, $x_2 \in V_0^{\perp}$. Then

$$V_0 = \{(x, 0) : x \in V_0\},$$

and it is possible to see this subspace as the graph of the zero operator acting from V_0 into $V_0^{\perp}$.

Lemma 7.1.4 *A subspace V lies in the open ball*

$$B(V_0, 1) = \{V : \varrho(V_0, V) < 1\}$$

if and only if it is the graph of a linear operator F, acting from V_0 into $V_0^\perp$, i.e.

$$V = \{(x_1, Fx_1) : x_1 \in V_0\} = \{x_1 + Fx_1 : x_1 \in V_0\}.$$

Proof Let F be an arbitrary linear operator F acting from V_0 into $V_0^\perp$. The graph of F is the set of pairs

$$\{(x_1, Fx_1) : x_1 \in V_0\}$$

and it is a d-dimensional subspace V in $\mathbb{C}^m$. Here $V \cap V_0^\perp = \{0\}$, which is equivalent to $\varrho(V_0, V) < 1$.

If $\varrho(V_0, V) < 1$, then the projector P_{V_0} bijectively maps V onto V_0 and, consequently, there exists the inverse operator $B_V : V_0 \to V$. Then each vector x from V may be represented in the form

$$x = (x_1, [I - P_{V_0}]B_V x_1), \quad x_1 \in V_0,$$

and we obtain the required operator

$$F_V = [I - P_{V_0}]B_V. \qquad\qquad \square$$

In this way, any subspace V, belonging to the open ball

$$B(V_0, 1) = \{V : \varrho(V_0, V) < 1\},$$

may be identified with an operator F_V and then the elements of the matrix of the operator F_V (in a fixed basis) of the format $d \times (m - d)$ are local coordinates in $G(m, d)$.

Let us construct transition functions from one of the coordinate systems in $G(m, d)$ to another. Let us look the coordinate systems in $G(m, d)$ related to the subspace V_1, and the coordinate system related to the subspace V_2.

Let $V \in B(V_1, 1) \cap B(V_2, 1)$ and the operator $F : V_1 \to V_1^\perp$ corresponds to the subspace V in the first coordinate system, i.e.

$$V = \{x = x_1 + Fx_1 : x_1 \in V_1\}.$$

Let us obtain the operator F_2, corresponding to V in the second coordinate system. For $x \in V$, we have the decomposition $x = y_1 + y_2$, where

$$y_1 = P(V_2)x \in V_2, \quad y_2 = [I - P(V_2)]x \in V_2^\perp.$$

Now, one has to find the operator F_2 such that $y_2 = F_2 y_1$. We have

$$y_1 = [P(V_2) + P(V_2)F]x_1, \ y_2 = \{[I - P(V_2)] + [I - P(V_2)]F\}x_1.$$

Since the projector $P(V_2)$ bijectively maps V onto V_2, on V_2 is defined the inverse operator for $P(V_2) + P(V_2)F$ and $x_1 = [P(V_2) + P(V_2)F]^{-1}y_1$. We obtain

$$y_2 = \{[I - P(V_2)] + [I - P(V_2)]F_1\}[P(V_2) + P(V_2)F_1]^{-1}y_1.$$

In this way, the transition from the old coordinate system to the new one is given by a matrix fractional linear map

$$\phi_A(F) = \{[I - P(V_2)] + [I - P(V_2)]F\}[P(V_2) + P(V_2)F]^{-1}.$$

Such a map is infinitely differentiable on its domain. Thus, it is shown that space $G(m, d)$ has a structure of infinitely differentiable manifold.

Let us sum up these considerations in the form of a theorem.

Theorem 7.1.5 *The space $G(m, d)$, consisting of d-dimensional subspaces of an m-dimensional complex space is compact and has the structrure of a $d(m - d)$-dimensional complex infinitely differentiable manifold.*

The set $G(m, d)$, with the introduced structure of a smooth manifold, is called the *Grassmann manifold*.

For $d = 1$, the manifold $G(m, 1)$ is the complex projective space CP^{m-1}, whose coordinates have already been introduced in the Sect. 5.4.

7.2 The Action of a Linear Map by Local Coordinates

The action of the linear non-singular operator $A : \mathbb{C}^m \to \mathbb{C}^m$ induces a bijective map $\phi_A : G(m, d) \to G(m, d)$. Let us write this map in a local coordinate system given on a neighborhood of the subspace V_0, i.e. we assume that $\varrho(V_0, V) < 1$.

Under the decomposition $\mathbb{C}^m = V_0 \oplus V_0^\perp$ a block matrix

$$A = \begin{pmatrix} A_{11} & A_{12} \\ A_{21} & A_{22} \end{pmatrix} \tag{7.2.1}$$

corresponds to the operator A. Since the subspace V may be represented in the form

$$V = \{x = \xi + F\xi : \xi \in V_0\},$$

we obtain that

$$A(V) = \{Ax = A_{11}\xi + A_{12}F\xi + A_{21}\xi + A_{22}F\xi : \xi \in V_0\}.$$

Here, the vector $\eta = A_{11}\xi + A_{12}F\xi$ belongs to V_0, and vector $A_{21}\xi + A_{22}F\xi$ belongs to $V_0^\perp$. Let us consider subspaces V for which $A(V)$ also belongs to the given neighborhood: $\varrho(V_0, A(V)) < 1$. Then, the operator $A_{11} + A_{12}F$ bijectively maps V_0 onto itself. Therefore,

$$\xi = (A_{11}\xi + A_{12}F)^{-1}\eta$$

and for an arbitrary vector from the subspace $A(V)$ we obtain the standard representation:

$$Ax = \eta + (A_{21} + A_{22}F)(A_{11} + A_{12}F)^{-1}\eta : \eta \in V_0.$$

In this way, the following result is obtained.

Lemma 7.2.1 *The action of the map ϕ_A in local coordinates (in the space of $d \times (m - d)$ matrices), is given by a fractional linear map of the form*

$$\phi_A(F) = (A_{21} + A_{22}F)(A_{11} + A_{12}F)^{-1}.$$

The map ϕ_A is called the *Möbius transformation*.

In Sect. 5.4 we have already discussed the case of the manifold $G(m, 1) = CP^{m-1}$ and describe the action of a linear operator on that manifold. Recall that the manifold $G(2, 1) = CP^1$ is the Riemann sphere, and the action of a linear map is given as an ordinary fractional linear transformation.

According to the previous discussion, the problem concerning the behavior of the trajectory of subspaces consists of the study of dynamics of nonlinear Möbius map. It is natural to expect that it is harder to obtain the description of the dynamics of a nonlinear map, than that of the linear. Let us clarify some technical difficulties.

In order to obtain an explicit expression for $\phi_{A^n}(F)$, one has to construct the decomposition of the form (7.2.1) for A^n, and then

$$\phi_{A^n}(F) = (A_{21}^n + A_{22}^n F)(A_{11}^n + A_{12}^n F)^{-1}.$$

We have obtained previously the representation of the matrix of the operator A in the Jordan basis. But, the decomposition (7.2.1) has to be constructed in another basis, determined by the subspace V_0 and, in addition to that, one has to construct the inverse matrix of the matrix $(A_{11}^n + A_{12}^n F)^{-1}$. Therefore, for $\phi_{A^n}(F)$ one gets complex expressions, inaccessible for direct study.

Let us give yet another remark. Invariant subspaces V are the only fixed points of the map ϕ_A. Therefore, matrices F corresponding to invariant subspaces, satisfy nonlinear equation

$$A_{21} + A_{22}F = F(A_{11} + A_{12}F).$$

This shows, from yet another point of view, that the set of invariant subspaces may have a rather complex structure.

Chapter 8
The Exterior Algebra and Subspaces

One of the approaches toward investigation of the Grassmann manifolds is based on an embedding of $G(m, d)$ into a projective space of large dimension, since this can be used to apply results on the action of a linear operator on a projective space. This embedding is achieved with the help of the Grassmann algebra.

8.1 The Exterior Algebra

Let X be a m-dimensional vector space with a given basis $e_1, e_2, \ldots, e_m$.

Definition 8.1.1 By an exterior algebra $\bigwedge X$ (Grassmann algebra) of the space X, one calls an associative algebra the unit, whose multiplication is denoted by $\wedge$, generated by elements $1, e_1, e_2, \cdots, e_m$ with determining relations

(1)
$$e_i \wedge e_j = -e_j \wedge e_i;$$

(2)
$$1 \wedge e_i = e_i;$$

(3)
$$1 \wedge 1 = 1.$$

These relations determine the algebra $\bigwedge X$ uniquely up to the natural isomorphism, and the algebra does not depend on the choice of the original basis.

Those elements which are of the form of the product od d basis vectors, i.e.

$$e_\gamma = e_{\gamma_1} \wedge e_{\gamma_2} \wedge \ldots \wedge e_{\gamma_d}, \tag{8.1.1}$$

© Springer International Publishing Switzerland 2014

Ć.B. Dolićanin and A.B. Antonevich, *Dynamical Systems Generated by Linear Maps*,
DOI: 10.1007/978-3-319-08228-8_8

where γ is a multiindex of length d:

$$\gamma = (\gamma_1, \gamma_2, \dots, \gamma_d), \ 1 \le \gamma_j \le m,$$

are called *elementary d-vectors*. Elements from $\bigwedge X$, which may be represented in the form of linear combination of elementary d-vectors, are called *d-vectors*.

The subspace of $\bigwedge X$, consisting of d-vectors, is denoted by $\overset{d}{\bigwedge} V$ and is called *exteriord power of the space X*.

The possible values are $d = 0, 1, 2, \dots, m$. In particular, the subspace $\overset{0}{\bigwedge} X$ is one-dimensional and consists of elements of the form $c\mathbf{1}, c \in \mathbb{C}$; the subspace of 1-vectors $\overset{1}{\bigwedge} X$ coincides with the original space X.

If in the product (8.1.1) the same basis vector occurs twice, this product became zero. If two such products differ only in the order of the factors, these products either coincide or differ only in sign. Therefore, as a basis of the space $\overset{d}{\bigwedge} X$ one may take products of the form (8.1.1), where the sequence of indices is strictly increasing:

$$\gamma_1 < \gamma_2 < \dots < \gamma_d.$$

The dimension of the space $\overset{d}{\bigwedge} X$ is equal to the number of such sequences, and it is equal to C_m^d—the number of the choices of d elements out of m. An arbitrary element from $\overset{d}{\bigwedge} X$ may be uniquely represented in the form

$$\sum_{\gamma \in \Gamma_d} a_\gamma e_\gamma,$$

where the sum is over the set of all multiindices satisfying the monotonicity condition:

$$\Gamma_d = \{\gamma_1 < \gamma_2 < \dots < \gamma_d \le m\}. \tag{8.1.2}$$

The Grassmann algebra may, as a vector space, be represented in the form of a direct sum of spaces:

$$\bigwedge X = \overset{m}{\underset{d=0}{\bigoplus}} \overset{d}{\bigwedge} X,$$

its dimension as a vector space is $2^m = \sum_d C_m^d$. In particular, the space $\overset{m}{\bigwedge} X$ is one-dimensional—it is generated by the element

$$e_1 \wedge e_2 \wedge \dots \wedge e_m.$$

The application of the exterior power $\overset{d}{\bigwedge} X$, for the study of d-dimensional subspaces, is based on the following statement.

Proposition 8.1.2 (The Grassmann Theorem). *A collection of linearly independent vectors*

$$v_1, v_2, \ldots, v_d$$

spans the same d-dimensional subspace, as a collection of vectors

$$w_1, w_2, \ldots, w_d,$$

if and only if the corresponding d-vectors

$$v_1 \wedge v_2 \wedge \ldots \wedge v_d \quad and \quad w_1 \wedge w_2 \wedge \ldots \wedge w_d$$

are proportional — they differ only by a scalar factor.

The Grassmann Theorem is one of the corollaries of the following fundamental theorem.

Theorem 8.1.3 *Let $\xi \in \overset{d}{\bigwedge} X$, $\xi \neq 0$. The set of vectors $x \in X$, such that*

$$\xi \wedge x = 0, \tag{8.1.3}$$

is a vector subspace V_ξ in X; two proportional vectors corresponds the same subspace.

If $v_1, v_2, \ldots, v_q$ is an arbitrary basis in V_ξ, then there is $(d - q)$-vector z, such that

$$\xi = v_1 \wedge v_2 \wedge \ldots \wedge v_q \wedge z.$$

Proof If we write the equality (8.1.3) in coordinates, we obtain a system of C_m^{d+1} linear homogeneous equations, with respect to m unknowns x_k—coordinates of vectors $x \in X$. This system may be written in the standard form

$$M(\xi)x = 0,$$

where the matrix $M(\xi)$ is constructed by the vector ξ. Therefore vectors $x \in X$, satisfying this system, form a vector subspace which we denote by V_ξ. The dimension of this subspace is $q = q(\xi) = m - r$, where r is the rank of the matrix $M(\xi)$. Obviously, for $C \neq 0$ we have

$$\xi \wedge x = 0 \Leftrightarrow C\xi \wedge x = 0,$$

i.e. the proportional vectors produce the same subspace.

Let us choose a basis $v_1, v_2, \ldots, v_m$ in the space X, such that the vectors $v_1, v_2, \ldots, v_{q(\xi)}$ form a basis in V_ξ. The vector ξ, as any d-vector, has the form

$$\xi = \sum_{\gamma \in \Gamma_d} a_\gamma v_\gamma, \tag{8.1.4}$$

where $v_\gamma = v_{\gamma_1} \wedge v_{\gamma_2} \wedge \ldots \wedge v_{\gamma_q}$. If the multi-index γ, does not contain the index i, then $v_\gamma \wedge v_i \neq 0$. Therefore, from the conditions $\xi \wedge v_i = 0$, for $i \leq q$, it follows that $a_\gamma = 0$ for all multi-indices γ, do not contain the set $\{1, 2, \ldots, q\}$. Hence, all vectors v_γ, appearing it the decomposition (8.1.4) with non-zere coefficients, contain as a factor q-vector $v_1 \wedge v_2 \wedge \ldots \wedge v_q$, and ξ is represented in the form

$$\xi = v_1 \wedge v_2 \wedge \ldots \wedge v_q \wedge z,$$

where z is some $(d - q)$-vector.   □

Corollary 8.1.4 *Let $q = q(\xi)$ be the dimension of the subspace V_ξ. For any non-zero d-vector ξ, one has the inequality*

$$q(\xi) \leq d.$$

The equality $q(\xi) = d$ holds if and only if the vector ξ is decomposable—it may be represented in the form of the product of 1-vectors:

$$\xi = v_1 \wedge v_2 \wedge \ldots \wedge v_d.$$

If $q(\xi) < d$, then $q(\xi) \leq d - 2$.
Accordingly, if r_ξ is the rank of the matrix $M(\xi)$, then $r_\xi \geq m - d$; the vector ξ is decomposable, if and only if $r_\xi = m - d$.

Proof If $q > d$, then ξ is represented in the form of the product of more than d vectors, i.e. this vector is not a d-vector.

For $q(\xi) = d$, the vector z is a 0-vector, i.e. a number, and from the theorem, we obtain the representation of the vector ξ in the form of the product of 1-vectors.

If $q = d - 1$, then the vector z is a 1-vector, satisfying the equality (8.1.3), then ξ may be represented as a product of 1-vectors and $q = d$. We get a contradiction.□

Corollary 8.1.5 *Let $v_1, v_2, \ldots, v_d$ and $v_1, v_2, \ldots, v_d$ be two linearly independent system of vectors in X. The subspaces V and U spanned respectively by these systems, coincide if and only if d-vectors $\xi = v_1 \wedge v_2 \wedge \ldots \wedge v_d$ and $\eta = v_1 \wedge v_2 \wedge \ldots \wedge v_d$ differ by a scalar factor:*

$$v_1 \wedge v_2 \wedge \ldots \wedge v_d = C v_1 \wedge v_2 \wedge \ldots \wedge v_d. \tag{8.1.5}$$

At the same time, if $B = \{b_{ki}\}$ is the transition matrix from the basis $\{v_i\}$ to the basis $\{v_k\}$, i.e.

$$v_k = \sum_k b_{ki} v_i,$$

then $C = \det B$.

Proof For the considered d vectors, we have $V_\xi = V$, $V_\eta = U$. If $V = U$, then, according to the Theorem 8.1.3, d-vector $\eta = v_1 \wedge v_2 \wedge \ldots \wedge v_d$ may be represented in the form

$$\eta = z \wedge v_1 \wedge v_2 \wedge \ldots \wedge v_q,$$

where z is some $(d-q)$-vector. In this case, $d - q = 0$, i.e., z – is a number, denoted by C.

A direct calculation shows that

$$v_1 \wedge v_2 \wedge \ldots \wedge v_d = \left(\sum_{k_1} b_{1k_1} v_{k_1} \right) \wedge \left(\sum_{k_2} b_{2k_2} v_{k_2} \right) \wedge \ldots \wedge \left(\sum_{k_d} b_{dk_d} v_{k_d} \right)$$
$$= \sum_{k_1} \sum_{k_2} \ldots \sum_{k_d} b_{1k_1} b_{1k_1} \ldots b_{dk_d} [v_{k_1} \wedge v_{k_2} \wedge \ldots \wedge v_{k_d}].$$

If some of the numbers k_i coincide, then the product $v_{k_1} \wedge v_{k_2} \wedge \ldots \wedge v_{k_d} = 0$. Thus, in the sum remains only summands corresponding to collections $k_1, k_2, \ldots, k_d$, where all numbers k_i are different. Each of such collection specifies a permutation $\mathbf{k}$ of indices $1, 2, \ldots, d$ and

$$v_{k_1} \wedge v_{k_2} \wedge \ldots \wedge v_{k_d} = (-1)^\sigma v_1 \wedge v_1 \wedge v_2 \wedge \ldots \wedge v_d,$$

where $\sigma = \sigma(\mathbf{k})$—the sign of the corresponding permutation. Therefore, equality (8.1.5) holds, where
$$C = \Sigma_{\mathbf{k}} (-1)^{\sigma(\mathbf{k})} b_{1k_1} b_{1k_1} \ldots b_{dk_d},$$

and the sum is taken over all permutations. And this quantity is exactly the determinant of the matrix B, being required.

We note in conclusion that, according to what has been proved, the d-vector corresponding to a d-dimensional subspace V, is determined by the given subspace uniquely up to the constant factor. Therefore, to non-proportional d-vectors there are corresponding different subspaces. $\qquad\square$

Note that we have used here the standard definition of the determinant of the matrix. In [1], the determinant of the matrix B is introduced as the coefficient of proportionality in the equality (8.1.5), and calculations, performed in the proof of the Corollary 8.1.5, represent an expression for the determinant being commonly taken in textbooks as its definition.

In some problems, the following interpretation of decomposable d-vectors turns out to be useful. Let $\mathbf{x} = (x_1, x_2, \ldots, x_d) \in \mathbb{C}^d$ be the coordinates of the vector $v \in V$ in the given basis $v_1, v_2, \ldots, v_d$:

$$v = \sum_k x_k v_k.$$

These coordinates determine on the space V a complex measure μ, invariant with respect to translations, generated by the standard complex measure on the space $\mathbb{C}^d$. This measure is uniquely defined by the equality for integrable functions

$$\int_V f(v)d\mu = \int_{\mathbb{C}^d} f(\mathbf{x})d\mathbf{x}.$$

Another basis $v_1, v_2, \ldots, v_d$ in V determines another coordinate system $v = \sum y_k v_k$, where these coordinate systems are related by the equations

$$x_i = \sum_k b_{ki} y_k,$$

i.e. $\mathbf{x} = B'\mathbf{y}$. For an integrable function, one has the formula of the change of variables in an integral:

$$\int_{\mathbb{C}^d} f(\mathbf{x})d\mathbf{x} = \int_{\mathbb{C}^d} f(B'\mathbf{y}) \det B \; d\mathbf{y}.$$

Therefore, the measures induced by these two bases, coincide, if and only if, $\det A = 1$. But, under the same conditions, d-vectors induced by these bases coincide:

$$v_1 \wedge v_2 \wedge \ldots \wedge v_d = v_1 \wedge v_2 \wedge \ldots \wedge v_d.$$

In this way, to each decomposable d-vector there corresponds a pair (V, μ), consisting of a d-dimensional subspace V and a (complex) measure μ on that subspace, invariant with respect to translations, and this correspondence is bijective.

8.2 An Embedding of the Grassmann Manifold into a Projective Space

Let us consider the map J, assigning to a d-dimensional subspace V one-dimensional subspace in $\overset{d}{\bigwedge} X$, spanned by the d-vector

$$\tilde{v} = v_1 \wedge v_2 \wedge \ldots \wedge v_d,$$

where $v_1, v_2, \ldots, v_d$ is a basis in V.

Recall that the space $\overset{d}{\bigwedge} X$ has dimension C_m^d, so the space of one-dimensional subspaces in it is a projective space $CP^{\tilde{d}}$, where $\tilde{d} = C_m^d - 1$. According to the previous discussion, the map J does not depend on the choice of a basis, and it determines an embedding of the Grassmann manifold $G(m, l)$ in the projective space $CP^{\tilde{d}}$. The map J is called the *Plücker embedding,* the coordinates of the vector $\tilde{v}$ are called the *Grassman coordinates* or the *Plücker coordinates* of the subspace V.

Note that map J is not surjective: its images are only decomposable d-vectors. Here, one has to pay attention to the fact that the set of decomposable d-vectors in $\overset{d}{\bigwedge} X$ is not a vector subspace. Moreover, dos not exist a linear relation, holds on this set. According to the Corollary 8.1.4, the conditions on the decomposability of a d-vector ξ consists in the requirement, that all minors of order $m - d + 1$ of the matrix M_ξ are equal zero. These minors are polynomials in coordinates of the vector ξ. This relations can be write as a quadratic equality, called *Plücker relationship.* Therefore, the Grassmann manifold embeds in a projective space as an algebraic subvariety.

In exceptional cases, the Grassmann manifold may coincide with the corresponding projective space. This by definition holds for $d = 1$ and $d = m - 1$. In particular, for $m = 3$, we have coincidence for all d. And in all other cases, there is no equality. This may be checked by comparison of dimensions: the Grassmann manifold has dimension $d(m - d)$, and the dimension $\tilde{d}$ of the corresponding projective space is the number

$$\tilde{d} = C_m^d - 1 = \frac{m(m - 1) \ldots (m - d + 1)}{d!} - 1.$$

Example 8.2.1 Let us clarify the description of the embedding J in the case of two-dimensional subspaces from the space X of dimension 4.

In this case, we have dim $G(4, 2) = 4$, the dimension of the corresponding projective space is $\tilde{d} = 5$ so, consequently, not all 2-vectors are decomposable. Let us find the condition on the decomposability of 2-vectors.

Suppose that e_1, e_2, e_3, e_4 is the basis in the space X. The space $X \bigwedge X$ has dimension 6, a basis in this space is formed by 2-vectors

$$w_1 = e_1 \wedge e_2, \ w_2 = e_1 \wedge e_3, \ w_3 = e_1 \wedge e_4,$$

$$w_4 = e_2 \wedge e_3, \ w_5 = e_2 \wedge e_4, \ w_6 = e_3 \wedge e_4.$$

For an arbitrary 2-vector

$$\xi = \sum_1^6 \xi_k w_k, \ \xi_k \in \mathbb{C},$$

let us consider the equation $\xi \wedge x = 0$, where

$$x = \sum_{1}^{4} x_i e_i \in X.$$

This equation may be rewritten in the form

$$\sum_{k=1}^{6} \sum_{i=1}^{4} \xi_k x_i \cdot w_k \wedge e_i = 0.$$

Here, 24 products $w_k \wedge e_i$ appear, some of these being equal to zero. Among non-zero products, there are 4 linearly independent. Considering this, we transform the equation:

$$(\xi_4 x_1 - \xi_2 x_2 + \xi_1 x_3) \, e_1 \wedge e_2 \wedge e_3 + (\xi_5 x_1 - \xi_3 x_2 + \xi_1 x_4) \, e_1 \wedge e_2 \wedge e_4$$
$$+ (\xi_6 x_1 - \xi_3 x_3 + \xi_2 x_4) \, e_1 \wedge e_3 \wedge e_4 + (\xi_6 x_2 - \xi_5 x_3 + \xi_4 x_4) \, e_2 \wedge e_3 \wedge e_4 = 0.$$

By equating to zero the coefficients of different 3-vectors, we obtain a linear system of four equations, with four unknowns x_k. The matrix of the system has the form

$$M(\xi) = \begin{pmatrix} \xi_4 & -\xi_2 & \xi_1 & 0 \\ \xi_5 & -\xi_3 & 0 & \xi_1 \\ \xi_6 & 0 & -\xi_3 & \xi_2 \\ 0 & \xi_6 & -\xi_5 & \xi_4 \end{pmatrix}. \tag{8.2.6}$$

According to the Corollary 8.1.4 the dimension $q(\xi)$ of the space of solutions of this system may only assume values 2 and 0. Hence, the rank $r(\xi)$ of the matrix $M(\xi)$ may only take values 2 and 4. This can be checked directly, using the explicit form of the matrix.

The condition on the decomposability of the 2-vector ξ is $r(\xi) = 2$. But, since the rank of matrices under consideration cannot be equal to 3, this is equivalent to the fact of the matrix being singular. In this way, the condition on the decomposability of a 2-vector ξ has the form

$$\det M(\xi) = 0,$$

where this determinant is a homogeneous polynomial of degree 4 in 6 variables.
    For example, 2-vector

$$\xi = w_3 + w_4 = e_1 \wedge e_4 + e_2 \wedge e_3$$

is indecomposable and $V(\xi) = \{0\}$. And 2-vector

$$\eta = w_1 + w_2 + w_3 + w_4 + w_5$$

is decomposable, the subspace $V(\eta)$ is two-dimensional and it is spanned by the vectors $v_1 = e_1 + e_2$ and $v_2 = e_2 + e_3 + e_4$.

Reference

1. Bourbaki, N. (1962). *Éléments de Mathématiques. Algebre*. Paris: Hermann.

Chapter 9
The Algebraic Approach to the Study of Subspace Trajectory

The algebraic approach is based on the previously established relations between a d-dimensional subspace V and the corresponding d-vector $\tilde{v} \in \overset{d}{\bigwedge} X$.

Exterior power of an operator and the action of model operators on $G(4, 2)$ will be more precisely described further on.

9.1 The Exterior Powers of an Operator

Let X be a m-dimensional vector space, V be a d-dimensional subspace of X. Let $v_1, v_2, \ldots, v_d$ be a basis in V. To subspace V, one associates d-vector $\tilde{v}$ from the space $\overset{d}{\bigwedge} X$, given by the expression

$$\tilde{v} = v_1 \wedge v_2 \wedge \cdots \wedge v_d. \tag{9.1.1}$$

Under the action of a non-singular linear operator A, the vectors $Av_k, k = 1, \ldots, d$, form a basis in the subspace $W = A(V)$. Thus, d-vector

$$\tilde{w} = Av_1 \wedge Av_2 \wedge \cdots \wedge Av_d.$$

corresponds to subspace $A(V)$. The right hand side of this formula determines an linear operator acting on the space $\overset{d}{\bigwedge} X$, called the *exterior power of the operator* A and denoted by $\overset{d}{\bigwedge} A$.

The operator $\overset{d}{\bigwedge} A$ acts on basis vectors $e_\gamma = e_{\gamma_1} \wedge e_{\gamma_2} \wedge \cdots \wedge e_{\gamma_d}$ by the rule

$$(\wedge^d A)e_\gamma = Ae_{\gamma_1} \wedge Ae_{\gamma_2} \wedge \cdots \wedge Ae_{\gamma_d}$$

and extends linearly to the whole space $\overset{d}{\bigwedge} X$.

© Springer International Publishing Switzerland 2014
Ć.B. Dolićanin and A.B. Antonevich, *Dynamical Systems Generated by Linear Maps*,
DOI: 10.1007/978-3-319-08228-8_9

From the presented discussion, the following proposition follows.

Proposition 9.1.1 *Let V be a d-dimensional subspace, A be a non-singular linear operator and $W = A(V)$. Then, the corresponding d-vectors $\tilde{v}$ and $\tilde{w}$ are related by equality*

$$\tilde{w} = (\bigwedge^{d} A)\tilde{v}.$$

In particular, the subspace V is invariant with respect to the operator A if and only if the corresponding d-vector $\tilde{v}$ is an eigenvector of the operator $\bigwedge^{d} A$.

This proposition is basic for the algebraic approach, consisting in the reduction of the question of the behavior of a trajectory of a subspace V in the space X, to the question on the behavior in $\bigwedge^{d} X$ of the trajectory of a one-dimensional subspace, spanned by the vector $\tilde{v}$. The operator $\bigwedge^{d} A$ is a linear operator in the space $\bigwedge^{d} X$ and for the description of the trajectory of a one-dimensional subspace one may apply the Theorem 5.4.2.

Let $\tilde{r}(k)$ be moduli of eigenvalus of the operator $\tilde{A} = \bigwedge^{d} A$. According to the Theorem 5.4.2, for each value $\tilde{r}(k)$ one distinguishes the subspace $M(k)$ in $\bigwedge^{d} X$, spanned by eigenvectors of modulus $\tilde{r}(k)$. The set of one-dimensional subspaces in $M(k)$ is a projective subspace P_k, embedded in the projective space $CP^{\tilde{d}}$.

In this way, the Grassmann manifold $G(m, d)$ is embedded in the projective space $CP^{\tilde{d}}$, and the behavior of the trajectory in the projective space is described in the Theorem 5.4.2.

By the Theorem 5.4.2, for any point $\xi \in CP^{\tilde{d}}$ can be constructed the number $k(\xi)$ and the point $\eta \in P_{k(\xi)}$, such that the trajectory of the point ξ tends to the trajectory of the point η. The trajectory of each point from P_k lies on a certain torus, and is described by the Proposition 5.2.1.

This leads to the following description of the behavior of the trajectory of subspaces.

Theorem 9.1.2 *Let A be a linear invertible operator acting on the m-dimensional complex space X. For given d-dimensional subspace V denote by $\tilde{v}$ the corresponding d-vector, given by (9.1.1) and denote by $\xi_V \in CP^{\tilde{d}}$ the one-dimensional subspace spanned by the vector $\tilde{v}$.*

There exists k, such that the trajectory of the subspace V converges to the sets

$$G_k = G(m, d) \bigcap P_k.$$

At the same time, there is, and may be constructed explicitly by using the previously listed formulas, the point $\eta \in G_k$ where the trajectory of the subspace V approaches to the trajectory of the subspace η.

In particular, a trajectory of a subspace V has the limit if and only if the corresponding subspace η is invariant.

The Theorem 9.1.2 solves theoretically the problem of the description of subspace trajectories. However, in the application of this theorem, one has to perform numerous complicated calculations. In particular, one has to study the operator $\bigwedge^d A$. One of the known general results on the properties of this operator is the following.

Proposition 9.1.3 [1] *Let $\lambda_1, \ldots, \lambda_m$ be all eigenvalues of the operator A, where each eigenvalue is written many times as is its multiplicity. Eigenvalues of the operator $\bigwedge^d A$ are all possible products of d eigenvalues of the operator A, i.e. the numbers of the form $\lambda_{i_1}\lambda_{i_2}\ldots\lambda_{i_d}$, where $i_1 < i_2 < \ldots < i_d$.*

In order to apply previous results, one has to know not only the eigenvalues of the operator $\bigwedge^d A$, but also the Jordan form of the operator. The general description of the Jordan form of the operator $\bigwedge^d A$ in not known to us, and fairly different situations arise in examples, requiring additional study.

Nevertheless, in some cases, one may get concrete results without complicated calculations.

Theorem 9.1.4 *Let the spectrum of the operator A consists of one point. Then, for each subspace V, there is the limit of the trajectory $A^n(V)$.*

Proof Since the operator A has only one eigenvalue λ, the operator $\bigwedge^d A$ has only one eigenvalue λ^d. Therefore, by the Corollary 5.4.3, there is the limit of the trajectory of each one-dimensional subspace under the action of the operator $\bigwedge^d A$, and, in particular, there is the limit of the trajectory of the one-dimensional subspace corresponding to V. This is equivalent to the existence of the limit of the trajectory $A^n(V)$.   □

In addition to technical complexities connected to the description of the Jordan form of the operator $\bigwedge^d A$, complications appear in the description of the possible forms of the behavior of trajectories of subspaces under the action of the given operator. For the operator $\bigwedge^d A$ (knowing its Jordan form), one may, by using the Theorem 5.4.2 obtain a description of all possible qualitatively different cases of the behavior of trajectories of one-dimensional subspaces in $\bigwedge^d X$. But, to the subspaces of V correspond only decomposable vectors, and examples show that for decomposable vectors one does not get all cases, possible for one-dimensional subspaces from $\bigwedge^d X$.

This may be shown in the example where for operator A all moduli of its eigenvalues λ_k are different. Here, it may happen that among products $\lambda_{i_1}\lambda_{i_2}\ldots\lambda_{i_d}$, there are those whose moduli are equal. Then, for the action of the operator $\bigwedge^d A$ on one-dimensional subspaces in $\bigwedge^d X$, there are trajectories having limit and trajectory with no limit. But, for decomposable vectors, only the first case occurs. Namely, the following results holds.

Theorem 9.1.5 *If moduli of different eigenvalues of the operator A are different, then for any subspace V there exist the limit of the trajectory $A^n(V)$.*

The proof of this theorem will be given below. Here, we only pay attention to the fact that for the proof of this theorem with the use of operator $\bigwedge^d A$, one has to perform fairly complex calculations; to construct the Jordan basis of the operator $\bigwedge^d A$, to record conditions on the decomposability of a d-vector by using the coordinates in this basis, and to obtain from these conditions the existence of the limit of the trajectory of the corresponding one-dimensional space.

9.2 The Actions of Model Operators on $G(4, 2)$

Let us discuss, as an example the problem, on the behavior of a trajectory of a two-dimensional subspace in a four-dimensional space, and use the example to clarify arising peculiarities.

Let us use the notation from Example 8.2.1. Suppose that e_1, e_2, e_3, e_4 is a basis in the space X. The space $X \wedge X$ has dimension 6 and 2-vectors

$$w_1 = e_1 \wedge e_2, \quad w_2 = e_1 \wedge e_3, \quad w_3 = e_1 \wedge e_4,$$

$$w_4 = e_2 \wedge e_3, w_5 = e_2 \wedge e_4, w_6 = e_3 \wedge e_4$$

form a basis in that space.

Recall that the condition on decomposability of a 2-vector ξ has the form

$$\det M(\xi) = 0,$$

where $M(\xi)$ is the matrix of the system (8.1.3).

Example 9.2.1 Let the operator A is given by one Jordan cell of dimension 4.

Without loss of generality, we may assume that the eigenvalue $\lambda = 1$ and the operator A acts by the formula

$$Ae_1 = e_1, Ae_j = e_j + e_{j-1} \text{ for } 2 \le j \le 4.$$

Consider the action of the operator $\tilde{A} = A \wedge A$, on the space $X \wedge X$. On basis vectors, we have

$$\tilde{A}w_1 = Ae_1 \wedge Ae_2 = e_1 \wedge (e_1 + e_2) = w_1$$

$$\tilde{A}w_2 = Ae_1 \wedge Ae_3 = e_1 \wedge (e_2 + e_3) = w_1 + w_2$$

$$\tilde{A}w_3 = Ae_1 \wedge Ae_4 = e_1 \wedge (e_3 + e_4) = w_2 + w_3$$

$$\tilde{A}w_4 = Ae_2 \wedge Ae_3 = (e_1 + e_2) \wedge (e_2 + e_3) = w_1 + w_2 + w_4$$

$$\tilde{A}w_5 = Ae_2 \wedge Ae_4 = (e_1 + e_2) \wedge (e_3 + e_4) = w_2 + w_3 + w_4 + w_5$$

$$\tilde{A}w_6 = Ae_3 \wedge Ae_4 = (e_2 + e_3) \wedge (e_3 + e_4) = w_4 + w_5 + w_6.$$

The matrix of this operator in the given basis has the form

$$\tilde{A} = \begin{pmatrix} 1 & 1 & 0 & 1 & 0 & 0 \\ 0 & 1 & 1 & 1 & 1 & 0 \\ 0 & 0 & 1 & 0 & 1 & 0 \\ 0 & 0 & 0 & 1 & 1 & 1 \\ 0 & 0 & 0 & 0 & 1 & 1 \\ 0 & 0 & 0 & 0 & 0 & 1 \end{pmatrix}.$$

It is obvious that all eigenvalues of this matrix are 1, but the matrix is not given in the Jordan form.

Let us construct the pyramid corresponding to this matrix, i.e. let us find eigenvectors and adjoint vectors of different orders. Consider a nilpotent operator $N = \tilde{A} - I$, given by the matrix

$$N = \begin{pmatrix} 0 & 1 & 0 & 1 & 0 & 0 \\ 0 & 0 & 1 & 1 & 1 & 0 \\ 0 & 0 & 0 & 0 & 1 & 0 \\ 0 & 0 & 0 & 0 & 1 & 1 \\ 0 & 0 & 0 & 0 & 0 & 1 \\ 0 & 0 & 0 & 0 & 0 & 0 \end{pmatrix}.$$

Then, the corresponding pyramid consists of subspaces of the form

$$L_1 = \ker N \subset L_2 = \ker N^2 \subset \dots.$$

Being easy to check, the subspace $L_1 = \ker N$, spanned by eigenvectors of the operator $\tilde{A}$ is two-dimensional and consists of vectors of the form $\xi = (\xi_1, \xi_2, \xi_2, -\xi_2, 0, 0)$; the vectors $(1, 0, 0, 0, 0, 0)$ and $(0, 1, 1, -1, 0, 0)$ form a basis in that space. Note that for vectors from L_1

$$\det M(\xi) = -\xi_2^4,$$

hence, among them, the decomposables are only vectors of the form $(\xi_1, 0, 0, 0, 0, 0)$.

Since dim $L_1 = 2$, the Jordan form of the matrix $\widetilde{A}$ consists of 2 Jordan cells. From the relation,

$$w_6 \xrightarrow{N} w_4 + w_5 \xrightarrow{N} w_1 + 2w_2 + w_3 + w_4 \xrightarrow{N} 3w_1 + 2w_2 \xrightarrow{N} 2w_1 \xrightarrow{N} 0$$

we see that the written vectors form a chain of adjoint vectors, and the vector w_6 is an adjoint vector of order 4. So, one of the Jordan cells has dimension 5 and the other has dimension 1.

Theorem 9.2.2 *Suppose that the operator A is given by one Jordan cell of dimension 4. Then in the space X, there is only one invariant subspace V_{12} of dimension 2 spanned by the vectors e_1 and e_2. A trajectory of every two-dimensional subspace V converges to this subspace.*

Note that for any one-dimensional subspace L of $X \wedge X$, there is the limit of the trajectory – one-dimensional subspace spanned by one of eigenvectors of the operator $\widetilde{A}$, and this limit may be different from the one-dimensional subspace L_{12}, spanned by the vector $e_1 \wedge e_2$. There is no contradiction with the theorem, since these subspaces L are spanned by indecomposable vectors, and the limit is subspace spanned by an indecomposable vector.

The 2-vector $\widetilde{v}$, corresponding to the two-dimensional subspace V, is decomposable, and for this vector the condition $\det M(\widetilde{v}) = 0$ holds. The set of decomposable vectors is closed and invariant with respect to the action of the operator $\widetilde{A}$. Therefore, the limit of the trajectory of the one-dimensional subspace, spanned by the vector $\widetilde{v}$ is also spanned by the decomposable eigenvector. As noted above, decomposable eigenvectors have the form $(\xi_1, 0, 0, 0, 0, 0)$, and to all of them, corresponds the same subspace L_{12}.

We remark that, in particular, indecomposable eigenvectors of the operator $\widetilde{A}$ appear in this example.

Example 9.2.3 Let operator A be given by two Jordan cells of dimension 2, i.e. A is given by the matrix (6.4.8).

In this case, there is an extensive family of invariant subspaces, described in the Example 6.6.1, and the problem consists of the search for the limit of a trajectory of an arbitrary two-dimensional subspace V. The application of previous arguments, in this case, leads to the simple answer.

Theorem 9.2.4 *Let the operator A in a four-dimensional space X be given by two Jordan cells of dimension 2, i.e. it acts by the formula*

$$Ae_1 = e_1, Ae_2 = e_2 + e_1 \ Ae_3 = e_3, Ae_4 = e_4 + e_3.$$

If the subspace V is invariant, then its trajectory is stationery and, obviously, converges to that subspace.

If the two-dimensional subspace V is not invariant, then its trajectory converges to the two-dimensional subspace V_{13}, spanned by basis vectors e_1 and e_3.

Proof Let us consider the action of the operator $\widetilde{A} = A \wedge A$, on the space $X \wedge X$. On basis vectors, we have

$$\widetilde{A}w_1 = Ae_1 \wedge Ae_2 = e_1 \wedge (e_1 + e_2) = w_1$$

$$\widetilde{A}w_2 = Ae_1 \wedge Ae_3 = e_1 \wedge (e_3) = w_2$$

$$\widetilde{A}w_3 = Ae_1 \wedge Ae_4 = e_1 \wedge (e_3 + e_4) = w_2 + w_3$$

$$\widetilde{A}w_4 = Ae_2 \wedge Ae_3 = (e_1 + e_2) \wedge (e_3) = w_2 + w_4$$

$$\widetilde{A}w_5 = Ae_2 \wedge Ae_4 = (e_1 + e_2) \wedge (e_3 + e_4) = w_2 + w_3 + w_4 + w_5$$

$$\widetilde{A}w_6 = Ae_3 \wedge Ae_4 = (e_3) \wedge (e_3 + e_4) = w_6.$$

The matrix of this operator, in the given basis has the form

$$\widetilde{A} = \begin{pmatrix} 1 & 0 & 0 & 0 & 0 & 0 \\ 0 & 1 & 1 & 1 & 1 & 0 \\ 0 & 0 & 1 & 0 & 1 & 0 \\ 0 & 0 & 0 & 1 & 1 & 0 \\ 0 & 0 & 0 & 0 & 1 & 0 \\ 0 & 0 & 0 & 0 & 0 & 1 \end{pmatrix}.$$

Two one-dimensional Jordan cells are evident here, corresponding to w_1 and w_6, and the subspace L, spanned by vectors w_2, w_3, w_4 and w_5, is invariant, where the vector w_2 belongs to this subspace and it is an eigenvector.

The nilpotent operator $N = \widetilde{A} - I$, in this case, is given by the matrix

$$N = \begin{pmatrix} 0 & 0 & 0 & 0 & 0 & 0 \\ 0 & 0 & 1 & 1 & 1 & 0 \\ 0 & 0 & 0 & 0 & 1 & 0 \\ 0 & 0 & 0 & 0 & 1 & 0 \\ 0 & 0 & 0 & 0 & 0 & 0 \\ 0 & 0 & 0 & 0 & 0 & 0 \end{pmatrix}.$$

Under the action of this operator, we have

$$w_5 \overset{N}{\to} w_2 + w_3 + w_4 \overset{N}{\to} 2w_2 \overset{N}{\to} 0.$$

This means that these vectors form the chain of adjoint vectors, and the vector w_5 is an adjoint vector of order 2. in addition, in the subspace L, there is yet another eigenvector $w_3 - w_4$. Thus, on the subspace L, the Jordan form of the operator consists of one cell of dimension 3, and one cell of dimension 1. Therefore, the basis where the operator $\widetilde{A}$ has the Jordan form, under the previously used special notation, consists of four eigenvectors

$$e(1) = w_1, \ e(2) = w_6, \ e(3) = w_3 - w_4, \ e(4, 1) = 2w_2$$

and two adjoint vectors

$$e(4, 2) = w_2 + w_3 + w_4, \quad e(4, 3) = w_5.$$

This allows to find the limit of the trajectory for an arbitrary two-dimensional subspace V, by using previously obtained formulas. Let $v(j, l)$ be the coordinates (in the constructed basis) of the 2-vector $\widetilde{v}$, corresponding to the subspace V. If $v(4, 2) = v(4, 3) = 0$, then the vector $\widetilde{v}$ is an eigenvector of the operator $\widetilde{A}$, and from this we obtain that the subspace V is invariant.

If at least one of the coordinates $v(4, 2)$, $v(4, 3)$ is different from zero, then the limit of the one-dimensional subspace spanned by $\widetilde{v}$ is the subspace spanned by the vector w_2. two-dimensional subspace V_{13} To this one-dimensional subspace in $X \wedge X$, two-dimensional subspace corresponding to vector w_2 is subspace V_{13} spanned by basis vectors e_1 and e_3 corresponds, being eigenvectors. In this way, the trajectory of every non-invariant subspace V converges to the same subspace V_{13}.   □

Reference

1. Glazman, I. M., & Lubich, Yu. I. (1969). *Finite-dimensional analysis problems*. Moscow: Nauka.

Chapter 10
The Geometric Approach

We consider the problem on the existence of the limit of a trajectory of a subspace V and the construction of this limit in an explicit form. In the algebraic approach to this problem, one performs formal calculations with the associated d-vector $\tilde{v}$, the behavior of the trajectory of this vector depends on the relation between its coordinates in the Jordan basis of the operator $\overset{d}{\bigwedge} A$. For example, the condition on the existence of the limit of the trajectory, may be written as the condition where some of the coordinates of the vector $\tilde{v}$ vanish. Such relations are certain conditions on the position of the subspace V, with respect to the pyramid (2.2.3) and (2.2.8), related to the original operator A. But, in the algebraic approach, it is difficult to clarify the geometric meaning of these conditions.

In addition to that, in the algebraic approach, one does not use the inner structure of the subspace V—its vectors, its subspaces, projections onto given subspaces related to the operator A, etc.

We consider some other approaches to this problem and from the beginning we discuss arising complexities.

10.1 Stable Bases

The most natural approach seems to be the following. We choose a basis $\{v_k\}$ in the space V, and follow the behavior of one-dimensional subspaces L_{kn}, spanned by vectors $A^n v_k$.

Recall that, with the given x, the existence of the limit of one-dimensional subspaces L_{xn} spanned by vectors $A^n x$, is equivalent to the fact that the sequence $A^n x$ may be renormalized (multiply by a numerical sequence γ_n), where the renormalized sequence $\gamma_n A^n x$ has a non-zero limit w.

First, it seems plausible that the condition on the existence of the limit V_0 of the trajectory $A^n(V)$ is equivalent to the existence of limits $\lim_{n \to \infty} L_{kn} = \tilde{L}_k$, and that the limit subspace V_0 is spanned by one-dimensional subspaces $\tilde{L}_k$.

© Springer International Publishing Switzerland 2014

Ć.B. Dolićanin and A.B. Antonevich, *Dynamical Systems Generated by Linear Maps*, DOI: 10.1007/978-3-319-08228-8_10

However, in reality, this is not true for two reasons.

First, it may turn out that there is the limit V_0 of the trajectory $A^n(V)$, but the sequence of one-dimensional subspaces L_{kn} is not convergent. Moreover, it may turn out that for any $x \in V$ the limit of the sequence of one-dimensional subspaces L_{xn} does not exist.

Second, it may happen that the limit V_0 of the trajectory $A^n(V)$ exists, there are limits $\tilde{L}_k$ of one-dimensional subspaces L_{kn}, but, at the same time, subspace $\tilde{L}_k$ are linearly dependent, and do not span the limit subspace V_0. In particular, the case when all subspace $\tilde{L}_k$ coincide may occur.

This fact reflects the following "bad" properties. Let $(L_1, L_2, \ldots, L_d)$, $L_k \in G(m, 1)$, be a set of one-dimensional subspaces, a let $[L_1, L_2, \ldots, L_d] \in G(m)$ be the subspace in X, spanning by this set. This correspondence generates a map

$$\pi : [G(m, 1)]^d \ni (L_1, L_2, \ldots, L_d) \to [L_1, L_2, \ldots, L_d] \in G(m). \qquad (10.1.1)$$

In the discussed case, we have d sequences L_{kn} of one-dimensional subspaces from X, for each k there is the limit

$$\lim_{n \to \infty} L_{kn} = L_{k0},$$

for the sequence of spanning subspaces $[L_{1n}, L_{2n}, \ldots, L_{dn}]$, there is the limit V_0. But, under map π, one cannot pass to the limit:

$$\lim[L_{1n}, L_{2n}, \ldots, L_{dn}] \neq [\lim L_{1n}, \lim L_{2n}, \ldots, \lim L_{dn}].$$

This means that the map π *is not continuous*, moreover *it is not closed.*

The complexity of the given problem is related to this peculiarity.

There are cases where the previous approach does give solution: the limits of L_{kn} exist and they spanned the required limit subspace.

A basis $\{v_k\}$, in the subspace V, is called *stable* if there are limits of one-dimensional subspaces

$$\lim_{n \to \infty} L_{kn} = \tilde{L}_k$$

and these limits are spanned by linearly independent vectors.

In this way, related to the general problem concerning construction of the limit of a trajectory of a subspace V, there arises a partial problem:
for what subspace V, there is at least one stable basis?

We note immediately that, if there are limit one-dimensional subspaces $\tilde{L}_k$, then they are spanned by eigenvectors $\tilde{v}_k$ of the operator A. If the basis in question is stable, then, by definition, the limit vectors $\tilde{v}_k$ are linearly independent. From this, we obtain a *necessary condition for the existence of a stable basis:*

Proposition 10.1.1 *If for a subspace V, there is a stable basis, then its dimension d does not exceed the number of the Jordan cells of the operator A.*

In this way, if the matrix of the operator has nontrivial Jordan cells, there is a subspace containing no stable basis. Finding the limit of a trajectory of a subspace in the case when there are no stable basis, turns out to be the most difficult, and its solution is one of the most essential results presented in the section.

10.2 The Perron Case: The Limit of Trajectory of a Subspace

In this subsection, we assume that the *Perron condition* is satisfied—all m eigenvalues of the operator A have different moduli. This is the case of general position: the complement to the set of such matrices in the group of all non-singular matrices has a dimension smaller than the dimension of the complete matrix group. This means that there are "less" matrices for which the Perron condition does not hold than those matrices where it holds. From the other point of view, this means that randomly selected matrix satisfies the Perron condition with probability 1.

It is natural to consider the class of matrices satisfying the Perron condition separately, since the discussion simplifies and the strongest propositions hold in this case. The basic result is that, in the Perron case, the limit of the trajectory exists for any subspace V and this limit may be found explicitly.

The behavior of one-dimensional subspaces in the Perron case is described in the Theorem 5.4.2 and the Corollary 5.4.3. In particular, the trajectory of any one-dimensional subspace has a limit. Therefore, if $(v_1, v_2, \ldots, v_d)$ is an arbitrary basis in V, then for any basis vector v_k the sequence of images $A^n v_k$ has, after the corresponding renormalization, the limit $\tilde{v}_k$. However, these limits do not determine uniquely the limit of the sequence of subspace $A^n(V)$. In particular, for subspace V in general position and a basis in general position, the limit of such a renormalized sequence for all v_k is an eigenvector v_m, corresponding to the eigenvalue of largest modulus. In other words, basis of general position is not stable.

Let us introduce objects being used in the formulation of the basic result of this subsection. As before, we assume that eigenvalues of the operator A are indexed in the order of increasing moduli: $|\lambda_1| < |\lambda_2| < \cdots < |\lambda_m|$, and the operator is given by a diagonal matrix in the basis consisting of the corresponding eigenvectors e_k.

The pyramid in the space X, related to the discussed operator, has the form

$$M_1 \subset M_2 \subset \cdots \subset M_m,$$

where $M(k)$ is the subspace spanned by the first k of the basis vectors $e_1, e_2, \ldots, e_k$. This pyramid is complete, and the indicator $k(x)$ is defined on it—the index of the level of the pyramid containing the vector x:

$$k(x) = \max\{j : x(j) \neq 0\}.$$

For the given subspace V, we denote by $K(V)$ the set of the values of the indicator $k(x)$ on V.

The set $K(V)$ contains d elements and it characterizes the position of the subspace V with respect to the pyramid M_k: the number k belongs to $K(V)$ if and only if the intersection of V and the corresponding level of the pyramid is non-empty: $V \bigcap \{M_k \setminus M_{k-1}\} \neq \emptyset$.

Let K be an arbitrary subset of the set $\{1, 2, \ldots, m\}$, consisting of d different elements, and let

$$W(K) = \{x = \sum_{k \in K} c_k e_k : c_k \in \mathbb{C}\}$$

be a d-dimensional subspace, spanned by eigenvectors e_k with indices k belonging to K. We remark that in the Perron case only these subspaces are invariant for the operator A.

Theorem 10.2.1 *If the Perron condition holds, then the limit of the trajectory $A^n(V)$ as $n \to +\infty$ exists for any subspace V and this limit is the subspace $W(K(V))$.*

In addition, an arbitrary subspace V, has a stable basis. The basis in a subspace V is stable, if and only if, it is normal with respect to the pyramid $\{V_k : k \in K(V)\}$, where $V_k = V \bigcap M_k$.

Proof Subspaces $V_k = V \bigcap M_k$ form a chain of nested subspaces in V, but some of them, may coincide. If from any of the group of coinciding subspaces we keep only one subspace having the smallest index, we obtain a complete pyramid of subspaces of V, containing only subspaces V_k for $k \in K(V)$. According to the Theorem 4.1.6, for such a pyramid there is a normal basis $\{v_k\}$, being essentially indexed by elements of the set $K(V)$. Here, $k(v_k) = k$ and, consequently, the set $K(V)$ is the set of values of the indicator $k(x)$ on that normal basis in V.

For any basis vector v_j, the limit of the sequence of the corresponding one-dimensional subspaces exists and it is the subspace spanned by the vector $e_{k(v_j)}$. These limit vectors are linearly independent if and only if the numbers $k(v_j)$ are different. And this is the characteristic property of the basis being normal with respect to the complete pyramid. $\square$

We remark that the property formulated in the theorem is characteristic: if $A \neq I$, then from the existence of a stable basis for any subspace, the fulfillment of the Perron conditions follows.

Corollary 10.2.2 *Let K be an arbitrary subset of the set $\{1, 2, \ldots, m\}$, consisting of d different numbers, and let $W(K)$ be a d-dimensional invariant subspace spanned by eigenvectors $e_k, k \in K$. The pool (attracting set) of the subspace $W(K)$ consists of those subspaces, where set of values of the indicator $k(x)$ coincides with the set K.*

Let us point out that the proof of the theorem by the algebraic approach is more complex. For such a proof, one has to check that for each one-dimensional subspace in $\bigwedge^{d} X$, spanned by a decomposable d-vector, there is the limit of the trajectory by the action of the operator $\bigwedge^{d} A$. The operator $\bigwedge^{d} A$ may have different eigenvalues with the same moduli, and in the space $\bigwedge^{d} X$ there are one-dimensional subspaces whose trajectory does not have the limit. Therefore, the proof requires several calculations involving conditions on the decomposability of d-vectors.

10.3 The Orthogonalization Process

Since it may turn out that a stable basis for a given subspace does not exist, different approaches are necessary for the solution of the problem in the general case. Let us discuss another approaches.

Geometric obstacle for a basis $\{v_k\}$ in V to be stable consists in the fact that the limits of the corresponding sequences of normalized vectors $\gamma_{nk} A^n v_k$ may be lineary dependent. It is possible to avoid this by the process of orthogonalization.

By the basis in the space $A^n(V)$, consisting of the vectors $A^n v_k$, it is possible to construct an orthonormal basis $\{e_k^n\}$ with the help of the standard Gram-Schmidt orthogonalization process. We then obtain a sequence of basis where the convergence to the same limit of different vectors is not possible—for any k, the distance from the vector $\{e_k^n\}$ to the subspace spanned by other vectors is always 1. Therefore, if there exists the limit e_k for any sequence e_k^n, then these limits are necessarily linearly independent, and determine a basis in the resulting limit space.

However, this approach has a series of shortcomings, and let us emphasize some of them.

1. It may turn out that limits

$$\lim_{n \to \infty} e_k^n$$

   do not exist, and the process of orthogonalization does not give a solution of determining the limit of the trajectory $A^n(V)$, even if this limit exists.
2. The process of orthogonalization is formally easy to describe, the elements e_k^n of the new basis are linear combinations of vectors $A^n v_k$. But, the expressions for vectors e_k^n are inconvenient for analysis: these expressions include the scalar products $\langle A^n v_i, A^n v_j \rangle$ and it is difficult to establish when the required limits do exist.
3. The orthogonalization construction does not consider account the position of the original subspace V, with respect to the pyramid, induced by the operator A. In addition, this construction is not canonical, since it depends on the choice of the original basis, and even on the numeration of the vectors of that basis.

Therefore, the approach based on the orthogonalization process can give explicit results only for some exceptional cases.

10.4 A Modified Approach

It is possible to modify the previous approach, by giving up on the construction of
an orthogonal basis. It is enough for each n to construct a basis $\{v_1^n, \ldots, v_d^n\}$, in the
space $A^n(V)$, in such a way that for each k, there is the limit

$$\lim_{n \to \infty} v_k^n = \tilde{v}_k$$

and that all limit vectors $\tilde{v}_k$ are linearly independent. Then, they shall form a basis
in the limit space.

   Suppose that $\{v_1, \ldots, v_d\}$ is an arbitrary basis in V. The transition from the basis
$\{A^n v_k\}$ to the basis $\{v_1^n, \ldots, v_d^n\}$ is determined by a non-singular operator D_n, acting
on the space $A^n(V)$. Therefore, the problem to construct desired bases $\{v_1^n, \ldots, v_d^n\}$
may be reformulated in the following way: to construct a sequence of linear operators
D_n such that the sequence $D_n A^n : V \to X$ has a limit Φ, being an operator from V
to X, and it is injective operator on V. Then, the subspace $\Phi(V) \subset X$ is the limit of
the trajectory $A^n(V)$.

   In this way, the problem of finding the limit of the trajectory $A^n(V)$ reduces to the
construction of a sequence of operators D_n, and finding the limit $\lim D_n A^n$. From the
technical point of view, it is more convenient to construct the sequence of operators
$B_n : V \to V$, where $D_n A^n = A^n B_n$.

   The case where a stable basis exists, fits into this scheme, in the one the corresponding operators B_n are given by diagonal matrices.

   The geometric approach consists in the construction of the required sequence of
operators B_n, starting from an analysis of the asymptotic behavior of the trajectory
of vectors from V. In this construction, the geometry of the problem is reconsidered,
in particular, properties of the operator A, and the position of the subspace V, with
respect to the pyramid induced by A, being essential.

Chapter 11
The Geometric Approach: An Operator with One Point Spectrum

Considering that the most difficult at explicit construction of subspace trajectory limits is the case when spectrum operator A consists of an item with more Jordan cells of various dimensions and at this point the emphasis will be on formulation and proof of basis theorem and asymptotic decomposition of sequences $A^n(x)$ i $A^n B_n{}^1$.

11.1 Formulation of the Basic Theorem and the Idea of the Proof

Let the spectrum of the operator A be one point λ, but the operator has many Jordan cells of different dimensions. The existence of the limit $\widetilde{V}$ of the trajectory $A^n(V)$ for this case is proved in the Theorem 9.1.4. The previous analysis of the problem shows that main difficulties in the *explicit construction of the limit* of the trajectory $A^n(V)$ arise exactly in this case.

The results obtained below, we state as a theorem.

Theorem 11.1.1 *Let the spectrum of the operator A consists of one point. Then, for any subspace V, one may construct by standard linear algebra operations, such as a linear operator $\Phi : V \to X$, so that*

$$\lim_{n\to\infty} A^n(V) = \Phi(V).$$

In the statement of the Theorem, it is essential that under the construction of the operator Φ, one uses only standard operations from linear algebra, such as the construction of the kernel of a linear operator, or construction of a complement to the given subspace. All limits appearing in the intermediate transformations are computed in an explicit form.

Let us clarify the idea of the proof given below. The sequence of subspaces $V_n = A^n(V)$, may be given in many different ways. In particular, for any bijective linear operator

$$B_n : V \to V$$

© Springer International Publishing Switzerland 2014

Ć.B. Dolićanin and A.B. Antonevich, *Dynamical Systems Generated by Linear Maps,*
DOI: 10.1007/978-3-319-08228-8_11

we have $V_n = A^n B_n(V)$, where the product $A^n B_n$ is a linear operator from V to X. We solve the following problem: to construct a sequence of operators $B_n : V \to V$, such that there exists the limit

$$\Phi x = \lim_{n \to \infty} A^n B_n x, \quad x \in V,$$

and the linear operator Φ maps V to X injectively. This is equivalent to the claim that for any non-zero element $x \in V$, the limit $\lim_{n \to \infty} A^n B_n x$ exists and it is different from zero.

The required operators B_n are constructed in several steps. As we have already seen, in the study of asymptotics of trajectories of points, graded linear operator arise. For example, the map W given by (4.4.6), associating to a point x the vector giving the direction of the principal term of the asymptotic of the trajectory, is a graded linear operator. We construct B_n^1 in a such way, that the principal term of $A^n B_n^1$ is the linear operator associated to W, i.e. it is the assembly of the corresponding linear operators. Then, for arbitrary $x \in V$, there is the limit $\lim_{n \to \infty} A^n B_n^1 x$. But, it may happen that $\lim_{n \to \infty} A^n B_n^1 x = 0$ for some x.

In the analysis of the behavior of the sequence $A^n B_n^1 x$, new graded linear operator arises, and we construct B_n^2 in an analogous way. As a result, we obtain the set of x, where $\lim_{n \to \infty} A^n B_n^2 x = 0$, is smaller than in the first step.

In finitely many steps, one constructs such a sequence B_n, that for any non-zero element $x \in V$, the limit $\lim_{n \to \infty} A^n B_n x$ exists and it is different from zero.

One may look at the suggested construction from another point of view. Suppose that the required sequence of operators B_n is constructed, and let $\{v_1, \ldots, v_d\}$ be an arbitrary basis in V. Then, for each n, vectors $A^n B_n v_k$ form a basis in the space $A^n(V)$, and for each k there is the limit

$$\lim_{n \to \infty} A^n B_n v_k = \Phi v_k,$$

the limit vectors Φv_k are linearly independent, and they form a basis in the limit subspace $\Phi(V)$. In such a way, the constructions described below are in fact equivalent to the construction of special bases in spaces $A^n(V)$.

We remark that the method described below is dictated by the asymptotic behavior of a trajectory, it is related to the structure of the operator, and to the position of the subspace in question. This indicates certain advantage of the suggested geometric method over the orthogonalization method.

11.2 Notation and Preliminary Facts

We use the notation and results from Sect. 6.3.

Without loss of generality we may assume that $\lambda = 1$ and that $A = I + N$, where N is a nilpotent operator. We consider the Jordan basis $e(i, l)$, indexed by two indices, where i is the number of the Jordan cell, $1 \le i \le q$, and the index l shows the order of the associated vector $e(i, l)$. Here, $1 \leq l \leq q_i$, where q_i is the dimension of the Jordan cell with number i.

The operator N acts by the formulas

$$Ne(i, l) = e(i, l - 1), \quad \text{if } l > 1;$$

$$Ne(i, 1) = 0.$$

We have $N^p = 0$, where $p = \max q_i$.

Let $M(l)$, where $l \leq p$, is the subspace spanned by vectors $e(i, l)$ with the given l. We denote by $P(l)$ the projector onto that subspace.

If $1 < l \leq p$, then the operator N injectively maps the subspace $M(l)$ into $M(l - 1)$, $N(M(1)) = 0$.

Subspaces

$$M^s = \bigoplus_{l=1}^{s} M(l), \quad s = 1, 2, \ldots p, \tag{11.2.1}$$

form a pyramid, induced by the operator A. To this operator we associate an indicator given by the formula $s(x) = s$, if $x \in M^s \backslash M^{s-1}$. This indicator was already mentioned in the description of the vector trajectory.

For an arbitrary subspace V, intersections

$$V_s = V \bigcap M^s$$

form a linearly ordered collection of subspaces in V. According to the Lemma 6.1.1, there is a decomposition

$$V = \bigoplus_{1}^{p} W(s), \tag{11.2.2}$$

where $W(s)$ is a subspace of V_s such that

$$W(s) \bigcap M^{s-1} = \{0\}.$$

If we denote by $S(V)$ the set of values of indicators $s(x)$ on V, it is not difficult to check that $W(s) \neq 0$ if and only if $s \in S(V)$, and that if $0 \neq x \in W(s)$, then $s(x) = s$.

Let $Q(s)$ be the projector on the subspace $W(s)$, acting on V. Then

$$x = \sum_{s} Q(k)s. \tag{11.2.3}$$

For the following, it is essential that since the space V is decomposed into a direct sum, an arbitrary linear combination of vectors $w(s) = Q(s)x$ also belongs to V.

For comparison, we note that, if one considers the decomposition

$$x = \sum_l P(l)x,$$

where $P(l)$ is the projector onto the subspace $M(l)$, then a linear combination of vectors $P(l)x$ does not have to belong to V.

In problems, related to the behavior of subspace trajectories, not only the principal terms of the asymptotic expansion of vector trajectories, but also other terms of the asymptotic expansion, obtained in the Theorem 3.3.1, are important. For the case under consideration, we have $A = I + N$, and the asymptotic expansion look simpler than in the general case. Let us present such a complete expansion using the introduced notations.

Lemma 11.2.1 *Let $A = I + N$. For the trajectory of a vector $x \in X$, there is a representation*

$$A^n x = \sum_{s=0}^{s(x)-1} C_n^s R(s)x, \qquad (11.2.4)$$

where operators $R(s)$ are given by the formula

$$R(s) = N^s \sum_{s<j} P(j)x,$$

where all summands are non-zero, the rate of growth of summands increases with the growth of s, and the highest rate of growth has the summand for $s = s(x) - 1$.

Proof Since $N^p = 0$, for $n > p$ by the Newton binomial formula we have

$$A^n = (I + N)^n = I + nN + C_n^2 N^2 + \ldots + C_n^{p-1} N^{p-1},$$

where C_n^s are binomial coefficients. Since $N^s M(l) = 0$ for $l \leq s$, for arbitrary $x \in X$, we have $N^s x = N^s \sum_{j>s} P(j)x$. Moreover, by the definition of the number $s(x)$ we obtain that $N^s x = 0$ for $s \geq s(x)$. This gives (11.2.4). It remains to note that with the growth of n, the binomial coefficient C_n^s behaves as n^s and that all summands in (11.2.4) are non-zero.   □

We proceed to the description of the corresponding construction.

11.3 The Asymptotic Expansion of the Sequence $A^n x$ and the First Renormalization

For the vectors from the set $M(l)\backslash M(l-1)$, $l > 1$, we have $s(x) = l$, therefore for such x expansion (11.2.4) has the form

$$A^n x = \sum_{s=0}^{l-1} C_n^s R(s, l) x, \qquad (11.3.5)$$

where

$$R(s, l) = N^s \sum_{s < j \le l} P(j) x.$$

If one divides the obtained expression by the leading coefficient C_n^{l-1}, then for the renormalized trajectory

$$[C_n^{l-1}]^{-1} A^n x = [C_n^{l-1}]^{-1} \sum_{s=0}^{l-1} C_n^s R(s, l) x$$

there is the limit, and this limit is the vector $R(l - 1, l) x$. In this way, one has a map Ω, given by the formula

$$\Omega x = R(l - 1, l) x = N^{l-1} P(l) x, \quad \text{if} \ \ x \in M(l) \backslash M(l - 1), \ 1 \le l \le p.$$

This map is, in particular, defined on the subspace V. By the construction, all vectors from the image $\Omega(V)$ belong to the limit subspaces $\widetilde{V}$, so the map Ω allows the construction of a part of the required subspace.

The constructed map Ω is nonlinear, and computations using this map become more complicated, in particular, in these computations one cannot use the theory of linear operators. But, the map Ω is graded-linear: the action of Ω on the set $M(l) \backslash M(l-1)$ coincides with the action of the linear operator $R(l-1, l)$. Therefore, by using the linear operator associated to Ω, one may simplify the analysis.

These ideas are realized as follows. For $x \in V$, we have a representation $x = \sum_l Q(l) x$, where $s(w(l)) = l$ for each non-zero vector $w(l) = Q(l) x$. Therefore, from (11.3.5), we obtain

$$A^n w(l) = \sum_{s=0}^{l-1} C_n^s R(s, l) w(l). \qquad (11.3.6)$$

The expressions obtained with different l are independent, and they have different rate of growth.

By the construction of subspaces $W(l)$, the projector $P(l)$ onto the subspace $M(l)$ injectively acts on this subspace. Therefore, the condition $w(l) \ne 0$ is equivalent to the condition $P(l) w(l) \ne 0$, and it is equivalent to the condition $R(l-1, l) w(l) \ne 0$. Note that projections $P(j) w(l)$ may be non-zero for $j < l$.

Let us perform a renormalization, by dividing each expression (11.3.6) by the leading coefficient C_n^{l-1}:

$$A^n[C_n^{l-1}]^{-1}w(l) = \sum_{s=0}^{l-1}[C_n^{l-1}]^{-1}C_n^s R(s, l)w(l).$$

Note that the leading term, in the obtained expression, is $R(l-1, l)w(l)$:

$$A^n[C_n^{l-1}]^{-1}w(l) = R(l-1, l)w(l) + o(1).$$

After the summation of expressions, we obtain

$$A^n \sum_{l=1}^{p}[C_n^{l-1}]^{-1}Q(l)x = \sum_{l=1}^{p}\sum_{s=0}^{l-1} C_n^s[C_n^{l-1}]^{-1} R(s, l)w(l). \qquad (11.3.7)$$

The performed renormalization is equivalent to the application to the vector $x \in V$ linear operators

$$B_n^1 = \sum_{l=1}^{p}[C_n^{l-1}]^{-1}Q(l),$$

i.e.

$$A^n \sum_{l=1}^{p}[C_n^{l-1}]^{-1}Q(l)x = A^n B_n^1 x.$$

We remark that the operators B_n^1 are invertible in V.

For the given $x \in V$, the principal term in the obtained expansion (11.3.7) is $\Psi_p^1 x$, where linear operator Ψ_p^1 is given by the formula

$$\Psi_p^1 = \sum_{l=1}^{p} R(l-1, l)Q(l).$$

In this way, we have

$$A^n B_n^1 x = \Psi_{1p}x + o(1).$$

Note that the linear operator Ψ_p^1, determining the principal term of the renormalized expression, is the operator associated to the graded-linear operator Ω, and it is an assembly of operators $R(l-1, l)$.

In this way, as a result of the first renormalization, the linear operator Ψ_p^1 has been constructed such that each vector from the image $\Psi_p^1(V)$ is a limit of some sequence

$$x_n = A^n B_n^1 x \in A^n(V).$$

Therefore, the subspace $\Psi_p^1(V)$ belongs to the limit subspace.

Then, there are two possibilities.

If ker $\Psi_p^1 = 0$, then dim $\Psi_p^1(V) = $ dim V, therefore lim $A^n(V) = \Psi_p^1(V)$ and the problem is solved—the required operator $\Phi = \Psi_p^1$.

However, it may turn out that ker $\Psi_p^1 \neq 0$. This can be seen from the following consideration. By the construction of operators $R(s, l)$, all images $R(l - 1, l)(W(l))$ belong to the subspace $M(1)$, therefore the image $\Psi_p^1(V)$ also belongs to the subspace $M(1)$. As it was clarified before, not all invariant subspaces belong to the subspace $M(1)$. Therefore, if the limit (invariant) subspace $\tilde{V}$ does not belong to the subspace $M(1)$, then such a subspace cannot be constructed as a result of the first renormalization. So, the condition ker $\Psi_{1p} = 0$ may hold only for subspaces V of the special form, and in general, there are subspaces for which ker $\Psi_p^1 \neq 0$.

Let us clarify, from another point of vies, what has been obtained as a result of the performed constructions.

Let the kernel of the operator Ψ_p^1 has dimension ν. We choose a basis $v_1, \ldots, v_d$ in the space V, where the vectors $v_1, \ldots, v_\nu$ form a basis in the space ker Ψ_p^1. Then, the vectors $\Psi_p^1(v_{\nu+1}), \ldots, \Psi_p^1(v_d)$ form a basis in the subspace $\Psi_p^1(V)$. In this way, the first renormalization allows the construction of some collection of linearly independent vectors in the limit space $\tilde{V}$—a part of the basis in that subspace.

This basis in V induces a decomposition

$$V = \ker \Psi_p^1 \bigoplus V_1,$$

where the subspace V_1 is spanned by vectors $v_{\nu+1}, \ldots, v_d$. At the same time, an injective map from the subspace V_1 into the limit subspace $\tilde{V}$ has been constructed.

One has to remark here that if one applies an analogous procedure to the subspace ker Ψ_p^1, then the subspace in $\tilde{V}$, constructed by ker Ψ_p^1, turns out to be a subspace of $\Psi_p^1(V)$, and in this way it is not possible to construct a subspace $\tilde{V}$. Therefore, for construction of $\tilde{V}$ one, we perform further renormalizations, using the whole space V, not only ker Ψ_p^1.

11.4 The Asymptotic Expansion of the Sequence $A^n B_n^1 x$ and the Second Renormalization

If ker $\Psi_p^1 \neq 0$, then one has to perform second renormalization, whose form is dictated by the asymptotic expansion of the sequence $A^n B_n^1 x$.

Let us construct such an asymptotic expansion, placing all summands in (11.3.7) with respect to the rate of growth.

The behavior of the summand in (11.3.7), with the given l and s, is determined by the coefficient

$$C_n^s [C_n^{l-1}]^{-1} = \frac{(s + 1)(s + 2) \ldots (l - 1)}{(n - s)(n - s - 1) \ldots (n - l + 1)}, \qquad (11.4.8)$$

with behaves like $(1/n)^{l-s-1}$. In particular, for any constant ν, these summands for which $l - s - 1 = \nu$ (if these summands exist, the value ν shall be called *admissible*), have identical behavior. Therefore, the sum of summands corresponding to the given admissible value ν is natural to consider as a unique expression. For a given ν, we have $s = l - \nu - 1$, so the sum of summands having identical rate of growth, has the form

$$\sum_{l=1}^{p} C_n^{l-\nu-1} [C_n^{l-1}]^{-1} R(l - \nu - 1, l) w(l). \tag{11.4.9}$$

In particular, for $\nu = 0$, we obtain the previously written principal term whose numerical coefficients do not depend on n.

In the construction of the first renormalization, it was essential that in the expansion of a trajectory all summands have a form of some vector, not depending on n, multiplied by a numerical coefficient which does depend on n. In the constructed expression (11.4.9), we have a different dependence on n: this expression is a sum of different vectors with different coefficients $C_n^{l-\nu-1}[C_n^{l-1}]^{-1}$. These coefficients have the same rate of growth, but they are different coefficients.

Therefore, the expression (11.4.9) is poorly adjusted to the renormalization procedure: under the division by the normalizing factor, new coefficients turn out not to be constant, but some functions of the variable n.

So, we represent (11.3.7) in the form of the sum of several summands in such a way that each summand has a form of some vector, not depending on n, multiplied by a numerical coefficient, depending on n.

We point out the fact that for summands appearing in (11.3.7), one has $s < l$. To the value $s = l-1$, previously described principal terms of the expansion correspond, whose coefficients are 1. Therefore, one really has to transform only those summands for which $s < l - 1$.

Let us use the notation

$$(n)_p = \frac{n!}{(n - p)!} = n(n - 1) \ldots (n - p + 1).$$

Such polynomials on variable n appear in numerous calculations, $(n)_p$ is called the Pochhammer symbol. In particular,

$$C_n^s = \frac{(n)_n}{(s)_s (n)_s}.$$

We transform fractions (11.4.8), as rational functions of variable n, to the common denominator. We obtain a uniform representation of the considered coefficients:

$$C_n^s [C_n^{l-1}]^{-1} = (s+1)(s+2) \ldots (l-1) \frac{P_{sl}(n)}{P_{01}(n)} = (s+1)_{l-s-1} \frac{P_{sl}(n)}{P_{01}(n)}, \tag{11.4.10}$$

where $s < l - 1$, and

$$
\begin{aligned}
P_{sl}(n) &= (n)_s (n-l)_{p-l} = n(n-1)(n-2)\ldots(n-s+1) \\
&\quad \times (n-l)(n-l-1)\ldots(n-p+1).
\end{aligned}
$$

Note that here, in the definition of polynomials $P_{sl}(n)$, the quantity p is assumed to be constant. We also remark that

$$
P_{01}(n) \equiv P_{l-1,l}(n) = (n)_p.
$$

The polynomial $P_{s,l}(n)$, has degree $p - l + s + 1$, and it can be represented in the standard form

$$
P_{s,l}(n) = \sum_{k=0}^{p-l+s+1} S(s, l, k) n^k,
$$

where coefficients $S(k, l, j)$ may be found in an explicit form. In particular, the leading coefficient here is $S(s, l, p - l + s + 1)$, and we also have

$$
S(s, l, p - l + s + 1) = 1.
$$

We remark that coefficients of the polynomial $P_{01}(n)$ are *Stirling numbers* used in combinatorics, and coefficients $S(s, l, k)$ are called *modified Stirling numbers*.

As a result, we obtain the representation of the considered coefficients

$$
C_n^s [C_n^{l-1}]^{-1} = \frac{(s+1)(s+2)\ldots(l-1)}{P_{01}(n)} \sum_{k=0}^{p-l+s+1} S(s, l, k) n^k,
$$

and expansion

$$
\begin{aligned}
A^n \sum_{l=1}^{p} [C_n^{l-1}]^{-1} Q(l) x &= \sum_{l=1}^{p} R(l-1, l) w(l) \\
&\quad + \sum_{l=1}^{p} \sum_{s=0}^{l-2} \frac{(s+1)(s+2)\ldots(l-1)}{P_{01}(n)} \\
&\quad \times \sum_{k=1}^{p-l+s+1} S(s, l, k) n^k R(s, l) w(l).
\end{aligned}
$$

We change the order of summation in two inner sums:

$$
\sum_{s=0}^{l-2} \frac{(s+1)(s+2)\ldots(l-1)}{P_{01}(n)} \sum_{k=1}^{p-l+s+1} S(s, l, k) n^k R(s, l) w(l)
$$

$$
= \sum_{k=0}^{p} \frac{n^k}{P_{01}(n)} \sum_{s=(k-p+l-2)_+}^{l-2} (s+1)(s+2)\ldots(l-1)S(s,l,k)R(s,l)w(l),
$$

where $(k-p+l-2)_+ = \max\{k-p+l-2, 0\}$. From this

$$
A^n \sum_{l=1}^{p} [C_n^{l-1}]^{-1} Q(l)x = \Psi_{1,p}(x) + \sum_{l=1}^{p}\sum_{k=1}^{p} \frac{n^k}{P_{01}(n)}
$$

$$
\times \sum_{s=(k-p+l-2)_+}^{l-2} (s+1)(s+2)\ldots(l-1)S(s,l,k)R(s,l)w(l).
$$

We, now, change the order of summation in the first and second sum:

$$
A^n \sum_{l=1}^{p} [C_n^{l-1}]^{-1} Q(l)x = \Psi_{1,p}(x) + \sum_{k=1}^{p-1} \frac{n^k}{P_{01}(n)} \sum_{l=1}^{p}
$$

$$
\times \sum_{s=(k-p+l-2)_+}^{l-2} (s+1)(s+2)\ldots(l-1)S(s,l,k)R(s,l)w(l).
$$

As a result, we obtain the required expansion where the dependence on n is manifested only through numerical coefficients $\frac{n^k}{P_{0p}(n)}$, and all terms of the expansion have different orders:

$$
A^n B_n^1 x = \Psi_p^1(x) + \sum_{k=1}^{p-1} \frac{n^k}{P_{01}(n)} \Psi_k^1(x), \tag{11.4.11}
$$

where

$$
\Psi_k^1 = \sum_{l=1}^{p} \sum_{s=(k-p+l-2)_+}^{l-2} (s+1)(s+2)\ldots(l-1)S(s,l,k)R(s,l)Q(l). \tag{11.4.12}
$$

As can be seen from (11.4.12), here each of the constructed operators Ψ_k^1 is an assembly of some linear operators induced by the corresponding decomposition of the space V.

For $k = p - 1$, we have $S(l-2, l, p-1) = 1$, therefore

$$
\Psi_{p-1}^1 = \sum_{l=2}^{p} (l-1)R(l-2, l)Q(l).
$$

If in the decomposition of the subspace V, only one non-trivial summand $W(k_1) = V$ appear, then the operator, obtained as a result of the first renormalization is injective and the second renormalization is not necessary. Therefore, in the considered case, we have several non-trivial summands in the decomposition of the subspace V. This is essential for the following discussion. From the explicit form of the operator Ψ_{p-1}^1 it follows, that its image belongs to the subspace M^2. In addition, if at least one of the subspaces $W(l) \neq \{0\}$ for $l > 1$, then the image of the operator Ψ_{p-1}^1 does not belong to the subspace $M(1) = M^1$. In other words, some adjoint vectors of the operator A belong to the image of this operator.

Moreover, for any $l > 1$ we have

$$\Psi_{p-1}^1(W(l)) \bigcap M(1) = \{0\}.$$

Indeed, let $w(l) \in W(l)$. Then

$$\Psi_{p-1}^1 w(l) = (l-1) R(l-2, l) w(l).$$

If

$$\Psi_{p-1}^1 w(l) \in M(1),$$

then

$$P(2) \Psi_{p-1}^1 w(l) = 0.$$

By definition, we have

$$R(l-2, l) = N^{l-2} [P(l-1) w(l) + P(l) w(l)].$$

Therefore,

$$P(2) \Psi_{p-1}^1 w(l) = P(2)(l-1) [N^{l-2} P(l-1) w(l) + N^{l-2} P(l) w(l)].$$

Here,

$$N^{l-2} P(l-1) Q(l) w(l) \in M(1),$$

so

$$P(2) N^{l-2} P(l-1) w(l) = 0.$$

Vector $N^{l-2} P(l) w(l)$ belongs to the subspace $M(2)$. Therefore,

$$P(2)[N^{l-2} P(l) w(l)] = N^{l-2} P(l) w(l).$$

From this, we obtain that

$$N^{l-2}(l-1) P(l) w(l) = 0.$$

Here, $P(l)(W(l))$ is a subspace in $M(l)$, and the operator N^{l-2} injectively maps subspace $M(l)$ into $M(2)$. Therefore, $P(l)w(l) = 0$.

As noted before, the operator $P(l)$ injectively maps subspace $W(l)$ into $M(l)$. Hence, from $P(l)w(l) = 0$ it follows that $w(l) = 0$, being required.

For $k < p - 1$, we obtain operators Ψ_k^1, having somewhat more complex form. The most important here is that the image of the operator Ψ_k^1 belongs to the subspace $M(p - k + 1)$, and its intersection with the subspace $M(p - k)$ is zero.

We point out that we have again obtained the expansion with only finitely many summands. Previously, we used in the expansion the scale of quantities C_n^s, allowing us to obtain expansion as a finite sum. In the new expansion (11.4.11), the scale of quantities $\frac{n^s}{P_{0p}(n)}$ turned out to be essential, permitted to obtain an expansion as a finite sum.

As already noted, one may construct expansions with respect to different scales, for example, with respect to the simplest scale consisting of power functions n^s. However, the use of other scales leads to expansions having infinitely many terms, and to more complicated expressions for the terms of an expansion.

Let us proceed to the construction of the second renormalization. The summands in the expansion (11.4.11) are dependent, therefore, multiplication one of the summands with the normalizing factor may lead out of the space V. In order to correctly perform renormalization, we split (11.4.11) into independent summands. To do this, we proceed as before.

We consider subspaces

$$S(1, i) = \bigcap_{k=i}^{p} \ker \Psi_k^1. \tag{11.4.13}$$

They obviously form a chain of nested subspaces:

$$S(1, 1) \subset \ldots \subset S(1, p) \subset V = S_{p+1}.$$

First of all, note that $S(1, 1) = \{0\}$. Indeed, if

$$x \in \bigcap_{k=1}^{p} \ker \Psi_k^1,$$

then for such x from (11.4.11), we obtain that

$$A^n B_n^1 x = \Psi_p^1(x) + \sum_{k=1}^{p-1} \frac{n^k}{P_{0p}(n)} \Psi_k^1(x) = 0.$$

But the operator $A^n B_n^1$ acts on V injectively. Therefore, $x = 0$. Note that this reasoning is based on the finiteness of the expansion (11.4.11).

Let us construct the decomposition of the subspace V, corresponding to the introduced chain of subspaces:

$$V = \bigoplus_{i=1}^{p} W(1, i). \tag{11.4.14}$$

Here

$$W(1, i) \backslash 0 \subset \{S(1, i+1) \backslash S(1, i)\}.$$

By $Q(1, i)$, we denote the corresponding projector onto $W(1, i)$. Since $\Psi_k^1 Q(1, i)x = 0$ for $k > i$, we obtain for $i < p$

$$A^n B_n^1 Q(1, i)x = \sum_{k=1}^{i} \frac{n^k}{P_{01}(n)} \Psi_k^1 Q(1, i)x. \tag{11.4.15}$$

This means that the principal term of the asymptotic for vectors from $W(1, i)$ is

$$\frac{n^i}{P_{01}(n)} \Psi_i^1 Q(1, i)x$$

and it is given by using the operator Ψ_i^1.

We can now perform renormalization, since the summands of the form (11.4.15) are independent. Dividing (11.4.15) by the leading coefficient $\frac{n^i}{P_{01}(n)}$, we obtain for $i < p$

$$A^n B_n^1 \frac{P_{01}(n)}{n^i} Q(1, i)x = \sum_{k=1}^{i} \frac{1}{n^{i-k}} \Psi_k^1 Q(1, i)x.$$

For $i = p$, we have

$$A^n B_n^1 Q(1, p)x = \Psi_p^1 Q(1, p)x.$$

Adding the obtained expressions with different i, we further obtain

$$\sum_{i=1}^{p-1} A^n B_n^1 \frac{P_{01}(n)}{n^i} Q(1, i)x = \sum_{i=1}^{p} A^n B_n^1 \sum_{k=1}^{i} \frac{1}{n^{i-k}} \Psi_k^1 Q(1, i)x.$$

Such a renormalization of the summands is equivalent to the application of the linear operator (acting on V)

$$B_n^2 x = Q(1, p) + \sum_{i=1}^{p-1} \frac{P_{01}(n)}{n^i} Q(1, i).$$

As a result, we have an expansion with respect to degrees $1/n$:

$$A^n B_n^1 B_n^2 x = \Psi_p^1 Q(1, p) x + \sum_{i=1}^{p-1} \sum_{k=1}^{i} \frac{1}{n^{i-k}} \Psi_k^1 Q(1, i) x$$

$$= \sum_{i=1}^{p} \Psi_i^1 Q(1, i) x + \frac{1}{n} \sum_{i=1}^{p} \Psi_{i-1}^1 Q(1, i) x + \dots .$$

This expansion may be written in the form

$$A^n B_n^1 B_n^2 xx = \sum_{j=0}^{p} \frac{1}{n^j} \Psi_j^2(x), \qquad (11.4.16)$$

where

$$\Psi_j^2 x = \sum_{i=j+1}^{p} \Psi_{i-j}^1 Q(1, i) x.$$

Note that each of the constructed operators Ψ_j^2 is an assembly of operators Ψ_k^1. By selecting the principal terms of the constructed expansion, we obtain

$$A^n B_n^1 B_n^2 x = \Psi_0^2 x + O(\frac{1}{n}),$$

where

$$\Psi_0^2 = \sum_{i=1}^{p} \Psi_i^1 Q(1, i).$$

The subspace $\Psi_0^2(V)$ in $\widetilde{V}$, constructed as a result of the second renormalization, is broader than the subspace $\Psi_p^1(V)$ constructed previously. This may be obtained from the following reasons. The subspace $\Psi_p^1(V)$ consists of eigenvectors of the operator A and it belongs to the subspace $M(1)$. The image of the operator Ψ_0^2 is the linear span of subspaces $\Psi_k^1(W(1, k))$—images of operators Ψ_k^1. The image $\Psi_k^1(W(1, k))$, if not trivial, belongs to the subspace M^{p-k+1}, and does not belong to the subspace M^{p-k}.

Let k_0 be the largest of the values k, for which the subspace $\Psi_k^1(W(1, k))$ is non-trivial. Then, there is x, where in the sum

$$\Psi_0^2 x = \sum_{i=1}^{p} \Psi_i^1 Q(1, i) x$$

there is one non-zero summand, belonging to the subspace M^{p-k_0+1}, and the rest of the summands belong to subspaces with smaller indices. This summand cannot be destroyed after summation and it does not belong $\Psi_p^1(V)$. Therefore, the dimension of the subspace $\Psi_0^2(V)$ is larger than the dimension of the subspace $\Psi_p^1(V)$. In this way, as a result of second renormalization, one indeed obtains wider subspace.

Next, we proceed as above.

There are two possibilities. If $\ker \Psi_0^2 = 0$, then the problem is solved and $\tilde{V} = \Psi_0^2(V)$.

If $\ker \Psi_0^2 \neq 0$, one has to perform further renormalizations.

11.5 The Subsequent Renormalizations

If $\ker \Psi_0^2 \neq 0$, then the next renormalization is performed as follows. We consider subspaces

$$S(2, k) = \bigcap_{j=0}^{k} \ker \Psi_j^2.$$

They form a chain of nested subspaces:

$$\{0\} = S(2, p) \subset S(2, p-1) \subset S(2, 1) \subset S(2, 0) = \ker \Psi_0^2 \subset V.$$

The equality $S(2, p) = \{0\}$, as before, we obtain as a consequence of the finiteness of the expansion.

Consider a decomposition of the space V, corresponding to this chain:

$$V = \bigoplus_{k=0}^{p-1} W(2, k),$$

where $W(2, k)$ is a subspace in $S(2, k)$, so that

$$W(2, k) \bigcap S(2, k-1) = \{0\}.$$

The corresponding projector onto the subspace $W(2, k)$, we denote by $Q(2, k)$.

We perform the third renormalization by applying operators

$$B_n^3 = \sum_{k=0}^{p-1} n^k Q(2, k).$$

We obtain a new expansion

$$A^n B_n^1 B_n^2 B_n^3 x = \sum_{j=0}^{p-1} \frac{1}{n^j} \Psi_j^3(x),$$

where arising operators Ψ_j^3 are assemblies of the previous ones. In particular, the main part of the obtained expansion is determined by the operator

$$\Psi_0^3 = \sum_{j=0}^{p-1} \Psi_j^2 Q(2, j).$$

If $\ker \Psi_0^3 = 0$, the problem is solved and $\widetilde{V} = \Psi_0^3(V)$.

If $\ker \Psi_0^3 \neq 0$, we perform subsequent renormalizations and obtain expression of similar forms with explicitly given coefficients.

As a result, in finite number ν of steps, we obtain operator $\Psi_{\nu,0}$ with trivial kernel, and then $\widetilde{V} = \Psi_{\nu,0}(V)$.

These considerations conclude the proof of the theorem.

The number of renormalizations, necessary to obtain the limit of a trajectory of a subspace V, depends on the position of the subspace V with respect to the pyramid of subspaces M^s. As previously mentioned, the simplest is the case when in the expansion (11.2.2) only one non-zero summand appear, occuring when for some s one has

$$V \subset M^s, \quad V \bigcap M^{s-1} = 0.$$

Then, the operator Ψ_p^1, constructed as a result of the first renormalization, has zero kernel. Then, the limit subspace $\widetilde{V}$ belongs to the subspace $M(1)$, consisting of eigenvectors. However, this is possible only for subspaces of small dimensions, since in that case $\dim V \leq \dim M(s)$, $\dim V \leq \dim M(1)$.

For subspaces of dimensions large enough there are more than one non-zero summands in the expansion (11.2.2). The collection of dimensions d_s of subspaces $W(s)$ in (11.2.2) is only one of the characteristics of the position of the subspace V with respect to the pyramid of subspaces M^s. Other characteristics of the position of V with respect to constructed pyramids are also important in the construction of renormalizations. For example, the renormalization depends on the fact which of the projections $P(k)(W(l))$ are non-zero.

11.6 An Example: Renormalizations for the Model Operator

Let us consider the model operator (6.4.8) as an example. Two-dimensional invariant subspace for this operator have been described in Sect. 6.4, limits of trajectories of two-dimensional subspaces have been found by using algebraic approach in Theorem 9.2.4.

Operator (6.4.8) acts in the space $X = \mathbb{C}^4$ by the matrix

$$A = \begin{pmatrix} 1 & 1 & 0 & 0 \\ 0 & 1 & 0 & 0 \\ 0 & 0 & 1 & 1 \\ 0 & 0 & 0 & 1 \end{pmatrix},$$

having two Jordan cells. Here, the vectors $e_1 = (1, 0, 0, 0)$ and $e_3 = (0, 0, 1, 0)$ are eigenvectors, and vectors $e_2 = (0, 1, 0, 0)$ and $e_4 = (0, 0, 0, 1)$ are adjoint vectors.

The pyramid corresponding to this operator is $M^1 \subset M^2 = X$, where the subspace M^1 is spanned by eigenvectors e_1 and e_3.

Let V be an arbitrary two-dimensional subspace spanned by some linearly independent vectors y and z. The subspace V may be represented in the form (11.2.2), and in this case, this decomposition has the form

$$V = W(1) \bigoplus W(2), \tag{11.6.17}$$

where $W(1) = (V \bigcap M^1) \subset M^1$, $W(2) \bigcap M^1 = 0$. Depending on dimensions

$$d_1 = \dim W(1), \quad d_2 = \dim W(2) = 2 - d_1,$$

three different cases of the position of the subspace V are possible:

(1) $V \bigcap M(1) = 0$.
(2) $\dim[V \bigcap M(1)] = 1$.
(3) $\dim[V \bigcap M(1)] = 2$, i.e. $V = M(1)$.

The corresponding collections of dimensions (d_1, d_2) are $(0, 2)$, $(1, 1)$ and $(2, 0)$.

Case 1 holds, if the condition

$$z(3)y(4) - z(4)y(3) \neq 0$$

is satisfied. This is the general position case.

Case 2 holds, if the conditions

$$z(3)y(4) - z(4)y(3) = 0,$$
$$|y(3)| + |y(4)| + |z(3)| + |z(4)| \neq 0$$

are satisfied.

The subspaces satisfying these conditions form a submanifold of the the Grassmann manifold $G(4, 2)$, whose dimension is 3 and less than the dimension of Grassmann manifold $G(4, 2)$.

In this case, one may choose a basis whose first basis vector (we denote it also by y) is in the subspace $M(1)$, i.e. without loss of generality, we may assume that y has the form

$$y = (y(1), 0, y(2), 0).$$

Case 3 is specific—there is only one subspace $V = M(1)$ satisfying this condition. In this case, the subspace V is invariant and the limit of its trajectory coincides with the space itself.

For all of these cases, we apply previously described geometric approach for the construction of the limit of subspace trajectory.

(1) Since $V \cap M(1) = 0$, under the decomposition (11.6.17) of the subspace V, we have

$$W(1) = 0, \quad W(2) = V,$$

i.e. in the decomposition only one non-trivial subspace occurs $W(2) = V$.

The projection operator onto the subspace $M(2)$ acts on the vector $x = (x(1), x(2), x(3), x(4))$ by the formula

$$P(2)x = (0, 0, x(3), x(4)).$$

The projection of the subspace $P(2)W(2)$ onto the subspace $M(2)$ is a two-dimensional subspace, spanned by vectors

$$(0, 0, y(3), y(4)), \quad (0, 0, z(3), z(4)).$$

These vectors are linearly independent and $P(2)W(2) = M(2)$.

The projection operator onto the subspace $M(1)$, acts by the formula

$$P(1)x = (x(1), x(2), 0, 0).$$

Note that here, depending on the position of the subspace V, the dimension $\dim P(1)W(2)$ may be equal to 0, 1 or 2. Here, operators $R(s)$ act by formulas

$$R(2)x = (x(3), 0, x(4), 0)$$

$$R(1)x = x.$$

The decomposition of the trajectory of the vector x has the form

$$A^n x = n(x(3), 0, x(4), 0) + (x(1), x(2), x(3), x(4)).$$

Since in the decomposition of the space only one non-trivial subspace $W(2) = V$ occurs, the first renormalization here reduces the division of the vector by n:

$$B_{1n}x = \frac{1}{n}x.$$

After the first renormalization, we obtain

$$A^n B_n^1 x = (x(3), 0, x(4), 0) + \frac{1}{n}(x(1), x(2), x(3), x(4)).$$

Here, the main part of the obtained decomposition is given by the operator

$$\Phi_2^1 x = (x(3), 0, x(4), 0),$$

the image $\Phi_2^1(V)$ is a two-dimensional subspace, and as a result of the first renormalization we obtain the required limit subspace—this is the subspace M^1. Thus, the general position case turns out to be the simplest. As it can be seen from the performed calculations, in this case, there is the limit of the normalized trajectory for the first basis vector y and there is the limit of the normalized trajectory for the second basis vector z, and the limit vectors are linearly independent. In other words, in this case a stable basis exists, moreover, any basis in V is stable.

(2) We now consider the case when $\dim[V \cap M(1)] = 1$. In this case, we choose a basis where the first basis vector y lies in the subspace $M(1)$, i.e.

$$y = y(1, 1)e(1, 1) + y(1, 2)e(1, 2).$$

Here, the decomposition (11.6.17) may be constructed in the following way. The subspace $W(1)$ is chosen uniquely: $W(1) = V \cap M(1)$, and this is a one-dimensional subspace, spanned by the vector $y = (y(1), y(2), 0, 0)$.

As $W(2)$, we may take any complementary subspace; we take as $W(2)$ one-dimensional subspace, spanned by the vector $z = (z(1), z(2), z(3), z(4))$, where $|z(3)| + |z(4)| \neq 0$. Then the representation $x = w(1) + w(2)$, induced by the decomposition (11.6.17), has the form

$$x = c_1 y + c_2 z.$$

Therefore, decomposition of the trajectory of vector $x \in V$, in this case, has the form

$$A^n x = c_2 n(z(3), z(4), 0, 0) + (c_1 y(1) + c_2 z(1), c_1 y(2) + c_2 z(2), c_2 z(3), c_2 z(4)).$$

The first renormalization in the considered case consists in the application of the operator, acting by the formula

$$B_n^1 x = w(1) + \frac{1}{n}w(2).$$

After the renormalization, we obtain the expression

$$A^n B_n^1 x = \Phi_2^1 x + \frac{1}{n}\Phi_1^1 x,$$

where

$$\Phi_2^1 x = c_1(y(1), y(2), 0, 0) + c_2(z(3), z(4), 0, 0),$$

$$\Phi_1^1 x = c_2(z(1), z(2), z(3), z(4)).$$

Further, two subcases may occur.

(2a) The vectors $(y(1), y(2), 0, 0)$ and $(z(3), z(4), 0, 0)$ are linearly independent. Then, the image $\Phi_2^1(V)$ is a two-dimensional subspace $M(1)$, and this subspace is the limit of a trajectory of any subspace V, where this condition on linear independence holds.

In this case, a stable basis exists, each basis is not stable. For the stability of the basis, it is necessary and sufficient that one of the basis vectors belongs to the subspace $V(1) = V \cap M(1)$.

(2b) The vectors $(y(1), y(2), 0, 0)$ and $(z(3), z(4), 0, 0)$ are linearly dependent. Then, the operator Φ_2^1 has one-dimensional kernel, and the image $\Phi_2^1(V)$ is a one-dimensional subspace in $M(1)$ spanned by the vector y.

In this case, second renormalization is necessary. Construct a new decomposition

$$V = W(1, 1) \bigoplus W(1, 2)$$

of the subspace V, induced by the pyramid of the form (11.4.13). In our case this pyramid has the form $\ker \Phi_2^1 \subset V$. The subspace $\ker \Phi_2^1$ may be easily constructed explicitly: one of the vectors

$$w = z(3)y - y(1)z$$

or

$$w = z(4)y - y(2)z$$

non-zero, and it spans one-dimensional subspace $\ker \Phi_2^1$.

The subspace $W(1, 1)$ is chosen uniquely:

$$W(1, 1) = \ker \Phi_2^1.$$

As $W(1, 2)$, one may take any complementary subspace to V; since $y \notin \ker \Phi_2^1$, as $W(1, 2)$ we take one-dimensional subspace spanned by the vector y. Then, for an arbitrary vector x from V, we have a new decomposition

$$x = c_1 y + c_2 w,$$

inducing a new representation of the trajectory after the first renormalization:

$$A^n B_n^1 x = \Phi_2^1 [c_1 y + c_2 w] + \frac{1}{n} \Phi_1^1 [c_1 y + c_2 w]$$

$$= c_1 y + \frac{1}{n} [c_1 y + c_2 w].$$

The operators, determining the second renormalization, act by the formula

$$B_n^2 x = B_n^2 [c_1 y + c_2 w] = c_1 y + n c_2 w.$$

After the second renormalization, we obtain the expression

$$A^n B_n^1 B_n^2 x = \Phi_0^2 x + \frac{1}{n} \Phi_1^2 x,$$

where

$$\Phi_0^2 x = c_1 y + c_2 w = x,$$

$$\Phi_1^2 x = c_1 y.$$

In the case under consideration, we obtain that the operator Φ_0^2 acts on V as the identical operator, $\Phi_0^2(V) = V$, and the limit of the trajectory also coincides with V.

In the case (2b), the condition of linear dependence of vectors $(y(1), y(2), 0, 0)$ and $(z(3), z(4), 0, 0)$ is satisfied. But, this condition is equivalent to the fact that the subspace V is invariant. In this way, we have found that in the case (2b), the subspace is invariant and the limit of the trajectory coincides with the space V itself. However, for the construction of the operator Φ by the given procedure, the second renormalization was necessary.

(3) If $V = M(1)$, this subspace is invariant.

In this way, in the case of the considered operator A, the limit of the trajectory for arbitrary two-dimensional subspace V has been found. In particular, for any non-invariant V the limit of the trajectory is subspace $M(1)$, spanned by eigenvectors.

This result was obtained in the Theorem 9.2.4 by the algebraic approach. It was established that if $\tilde{v}$ is a 2-vector corresponding to the subspace V, then the form of the limit of the trajectory depends on the fact which of the coordinates $v(j, l)$ of that vector in the Jordan basis for operator $\tilde{A}$, are different from zero. If $v(4, 2) = v(4, 3) = 0$, then the vector $\tilde{v}$ is an eigenvector of the operator $\tilde{A}$, and the subspace V is invariant. If at least one of the coordinates $v(4, 2)$, $v(4, 3)$ is non-zero, then the limit of the one-dimensional subspace spanned by $\tilde{v}$ is the subspace, spanned by the vector w_2. To this subspace in $X \wedge X$ the subspace $V_{13} = M(1)$ corresponds, spanned by basis vectors e_1 and e_3.

Note, that the study with the help of geometric approach allows one to clarify the geometric meaning of formally obtained conditions, where one of the coordinates $v(4, 2)$, $v(4, 3)$ is non-zero. The condition $v(4, 3) \neq 0$ holds for subspace in general position, when the intersection of two-dimensional subspace $V \cap M(1) = 0$. In the geometric approach, this is the case (1). And the geometric meaning of the condition $v(4, 3) = 0$, $v(4, 2) \neq 0$ is that the case (2a) occurs, described in the geometric approach: the intersection $V \cap M(1)$ is a one-dimensional subspace and, at the same time, V is not invariant.

In particular, from these considerations, one may give the following observation: the construction of the limit of trajectory for the subspace V in a more general position is simpler. This is a manifestation of general laws to be established below.

11.7 The Typical Number of Renormalizations

The number of renormalizations, necessary for the construction of the limit of the trajectory of subspaces V, depends on the position of V with respect to the pyramid, induced by the operator A. Let us clarify what position of the subspace V is typical from the point of view of the number of renormalizations, necessary for the construction of the limit.

The precise meaning to the notion of typical is given as follows.

Let Y be a space and some property of points from Y is given. Suppose that a notion of *small subset* is introduced for subsets of Y. One says that the property *holds for almost all points* or *it is typical*, if the subset of points where the property does not hold is a small subset. The points where the property hold are called *points in general position* (with respect to the given property). Usually, as small subsets, one takes sets of measure zero, sets of first category in the sense of Baire or submanifolds of smaller dimension.

In the question under consideration, the original set is a manifold, and by small subsets we understand submanifolds of smaller dimension or unions of such manifolds; these subsets are also of measure zero, and are the sets of first category in the sense of Baire.

Related to the introduced notions, one also uses probability interpretation: the property is typical, if randomly selected point with probability 1 has that property.

Let us give a few examples of the typical properties.

Proposition 11.7.1 *In* m^2-*dimensional subspace* $\mathbb{C}^{m \times m}$, *consisting of square matrices of format* $m \times m$, *the property* $\det A \neq 0$ *is typical.*

Proof The set of points, where this property is not satisfied, is given by the equation $\det A = 0$ and it is a submanifold (with singularities) of dimension $m^2 - 1$. □

Proposition 11.7.2 *In* mn-*dimensional space* $\mathbb{C}^{m \times n}$, *consisting of rectangular matrices of format* $n \times m$, *the property where the rank of a matrix is maximum possible is typical:* $\operatorname{rank} A = \min\{n, m\}$.

Proof Suppose that $n \leq m$. The set of matrices where this property is not satisfied is given by the collection of equations $D(i_1, i_2, \ldots, i_n) = 0$, where $D(i_1, i_2, \ldots, i_k)$ is a determinant of the square matrix formed from the columns of the matrix A with indices $i_1, i_2, \ldots, i_n$. The number of such matrices (and corresponding equations) is C_m^n, therefore the condition $rank A = n$ is not satisfied only in intersection of finitely many submanifolds, whose dimensions are smaller than mn.   □

The question concerning typical properties arises in the description of the mutual position of a collection of subspaces.

If V and U are subspaces in a m-dimensional space X, where $\dim V = d$, $\dim U = k$, then the quantity $\dim(V \cap U)$ may assume values from $\max\{0, d + k - m\}$ to $\min\{d, k\}$.

Lemma 11.7.3 *For a pair of subspaces, the typical position is the one for which the quantity* $\dim(V \cap U)$ *has the smallest value:*

$$\dim(V \cap U) = \max\{0, d + k - m\}.$$

Proof The set of pairs of subspaces is the product $G(m, d) \times G(m, k)$, and the question is about a typical property of a point $(V, U) \in G(m, d) \times G(m, k)$.

Let us choose the basis $v_1, \ldots, v_d$ in V and $u_1, \ldots, u_k$ ine U. A vector x belongs to $V \cap U$, if there are constants c_i, $1 \le i \le d$ and constants c'_j, $1 \le i \le k$, so that

$$x = \sum_i c_i v_i = \sum_j c'_j u_j.$$

Form a matrix B having $d + k$ columns, taking as first d columns vectors v_i and as the last k columns, vectors u_j. Then the last equality is equivalent to the fact that the column vector c, formed from the numbers c_i and $-c'_j$, is the solution of the system $Bc = 0$.

According to the Proposition 11.7.2, the property that $rank B = \min\{d + k, m\}$ is typical. Therefore, if $d + k \le m$, in typical case the system $Bc = 0$ has only zero solution, and it follows that $V \cap U = 0$.

If $d+k > m$ the system $Bc = 0$ has, in typical case, $d+k-m$ linearly independent solutions, these solutions span the subspace $V \cap U$, following that its dimension is $d + k - m$.   □

The following Lemmas may be proved similarly as the Lemma 11.7.3.

Lemma 11.7.4 *Let us consider the set of collections of subspaces* W_k, $1 \le k \le p$ *in a finite-dimentional space* X.
If

$$\sum_{k=1}^{p} \dim W_k \le \dim X,$$

a typical property of such a collection is the verity of the equality

$$\dim(W_1 + W_2 + \ldots + W_l) = \sum_{k-1}^{p} \dim W_k,$$

which is equivalent to the equality

$$W_1 + W_2 + \ldots + W_p = \bigoplus_k W_k.$$

If

$$\sum_{k=1}^{p} \dim W_k \geq \dim X,$$

then the typical property of that collection is the verity of the equality

$$W_1 + W_2 + \ldots + W_p = X.$$

Lemma 11.7.5 *Let*

$$M^1 \subset M^2 \subset \ldots \subset M^p = X$$

be a pyramid of subspaces in a finite-dimentional space X, where $\dim M^k = \nu_k$. *For a subspace V of dimension d the typical is a position of V with respect to the given pyramid such that for all k the dimension of the intersection* $V \cap M^k$ *is minimal:*

$$\dim(V \cap M^k) = \max\{0, d + \nu_k - m\}.$$

Let us reformulate this proposition in terms of decomposition (11.2.2):

$$V = \bigoplus_1^p W(k),$$

where $\dim W(k) = d_k$. Denote the weights of the levels of the pyramid as $s_k = \nu_k - \nu_{k-1}$. The decomposition of a typically positioned subspace V is such that the dimension d_p has the largest of possible values $d_p = \min\{d, s_p\}$. In particular, if $d \leq s_p$, we have $d_p = d$ and $d_k = 0$ for $k < p$. In the case $d > s_p$, we have $d_p = s_p$ and d_{p-1} has the largest of possible values: $d_{p-1} = \min\{d - s_p, s_{p-1}\}$. Further, we have a similar pattern: if $d - s_p \leq s_{p-1}$, we have $d_{p-1} = d - s_p$ and $d_k = 0$ for $k < p - 1$; if $d - s_p > s_{p-1}$, we have $d_{p-1} = s_{p-1}$ and $d_{p-2} = \min\{d - s_p - s_{p-1}, s_{p-2}\}$.

Let us single out the case of the least typical posinion of V, opposite to the case of general position. In this case, d_1 reaches the largest possible value: $d_1 = \min\{d, s_1\}$, further d_2 reaches the highest possible value: $d_2 = \min\{d - d_1, s_2\}$ etc.

We have previously described a sequence of renormalizations leading to the construction of the limit $\widetilde{V}$ of the trajectory of a subspace V. It was demonstraited, that this construction depends on the position of V with respect to pyramid 11.2.1, induced by the operator A.

The concept of a typical position may be applied to the set of invariant subspaces and distinguish the most typical and the least typical ones with respect to pyramid 11.2.1.

For the operator from the model example, discussed in Sect. 11.6, it has been shown that the trajectory of a typical two-dimensional subspace V converges to the subspace $M(1)$, and here the subspace $M(1)$ is the least typical two-dimensional invariant subspace. This is a manifestation of the general pattern described in the following theorem.

Theorem 11.7.6 *Suppose that the conditions of the Theorem 11.1.1 are satisfied.*

If $d \leq \dim M(1)$, then operator Ψ_p^1, constructed as a result of first renormalization, is injective for almost all subspace V and it maps subspace V into $\widetilde{V}$—the limit of trajectory. At the same time, the limit subspace belongs to $M(1)$, and it is one of the least typical invariant subspace.

If $d > \dim M(1)$, then for almost all subspaces V operator Ψ_0^2, constructed as a result of the second renormalization, is injective and it maps subspace V into the limit of its trajectory. In addition to that, the limit subspace $\widetilde{V}$ belongs to M^s, where s can be found from the conditions $\dim M^{s-1} < d \leq \dim M^s$, and it is one of the least typical invariant subspace.

Proof Suppose that $d \leq \dim M(1)$. Consider the decomposition (11.2.2) of the subspace V. As follows from the construction of the operator Φ_p^1, the subspace $\Phi_p^1(V)$ is a sum of subspaces $R(l - 1, l)(W(l))$, where all of them belong to $M(1)$ and

$$\sum_l \dim[R(l - 1, l)(W(l))] = d.$$

According the to Lemma 11.7.4, the case when the sum of subspaces is direct and $\dim \Psi_p^1(V) = d$, is typical.

Suppose now, that $\dim M(1) < d$. Consider the decomposition (11.4.14), used in the construction of the second renormalization and the constructed operator

$$\Psi_0^2 = \sum_{i=1}^p \Psi_i^1 Q(1, i).$$

As noted in the construction of the second renormalization, the subspace $\Psi_0^2(V)$ is a sum of subspaces $\Psi_k^1(W(1, i))$, and the subspace $\Psi_i^1(W(1, i))$ belongs to the subspace M^{p-i+1}.

According to the Lemma 11.7.5 in the decomposition of typical subspaces, non-trivial are subspaces $W(1, i)$ with the highest indices, and then subspaces $\Psi_i^1(W(1, i))$ belong to subspaces M^k with small indices k. From this, we obtain, that for subspaces V in general position, all subspaces $\Psi_i^1(W(1, i))$ belong to M^s and, according to the Lemma 11.7.4, in the typical case the sum of subspaces $\Psi_k^1(W(1, i))$ is a direct sum.

In addition, the limit invariant subspace $\Psi_0^2(V)$ belongs to M^s, which is not true for typical invariant subspaces.   □

Let us illustrate this theorem in an example, discussed in the Sect. 11.6.

We first consider two-dimensional invariant subspaces. Here, a typical invariant subspace U is given by a vector

$$u = \sum_{ij} u(i, j) e(i, j)$$

from X, whose projection onto the subspace $M(2)$ is different from zero:

$$u(2, 1)e(2, 1) + u(2, 2)e(2, 2) \neq 0.$$

Here, U itself is a subspace spanned by vector u and vector

$$\begin{aligned} Au = &[u(1, 1) + u(2, 1)]e(1, 1) + [u(1, 2) + u(2, 2)]e(1, 2) \\ &+ u(2, 1)e(2, 1) + u(2, 2)e(2, 2). \end{aligned}$$

If the second projection

$$P(2)u = P(2)Au = u(2, 1)e(2, 1) + u(2, 2)e(2, 2)$$

is given, the first projection $P(1)Au$ may be an arbitrary vector from the space $M(1)$.

As shown before, the set $Inv(A) \subset G(4, 2)$ of such subspaces has a natural structure of a two-dimensional complex manifold. The subspace $M(1)$ is the least typical from the point of view of the position with respect to the pyramid $M(1) \subset X$, and all other invariant subspaces are subspaces in general position.

Let us consider all two-dimensional subspaces. Two-dimensional subspaces V of a four-dimensional space form the Grasssmann manifold $G(4, 2)$, is a four-domensional complex manifold. Typically from the point of view of position with respect to the pyramid $M(1) \subset X$ are subspaces such that $V \cap M(1) = 0$, i.e. the case (1) from 4.4. in this case, the limit of the trajectory is the subspace $M(1)$—the least typical invariant subspace.

In case (2) from 4.4, the condition

$$V \cap M(1) \neq 0,$$

is equivalent to the validity of the scalar equality

$$z(2, 1)y(2, 2) - z(2, 2)y(2, 1) = 0.$$

Therefore, the set of subspaces where this condition holds, form a three-dimensional subspace $W(3)$ in $G(4, 2)$.

Among subspaces belonging to $W(3)$, the condition that vectors $z(2, 1)e(1, 1) + z(2, 2)e(1, 2)$ and $y = y(1, 1)e(1, 1) + y(1, 2)e(1, 2)$ are linearly dependent (subcase 2b), form a submanifold $W(2)$ of dimension 2. As shown in the discussion of that example, all subspaces, being points from $W(2)$ are invariant subspaces.

In this way, we obtain partition of the Grassmann manifold $G(4, 2)$ into four disjoint subsets:

$$G(4, 2) = \{G(4, 2)\backslash W(3)\} \coprod \{W(3)\backslash W(2)\} \coprod \{W(2)\backslash M(1)\} \coprod M(1).$$

The points of the set $G(4, 2)\backslash W(3)$ are subspaces in general position, and the point $M(1)$ is the least typical invariant subspace. As presented in the example (being consistent with the Theorem), trajectory of each point from $G(4, 2)\backslash W(3)$ convergest to $M(1)$. Moreover, it was shown in the example, that trajectories of all points from $W(3)\backslash W(2)$ also converge to $M(1)$.

Let us emphasize another property of the considered example. Each subspace U from $W(2)\backslash M(1)$ is invariant, but there are no subspase V, $V \neq U$, whose trajectory converge to U.

Chapter 12
The Convergence of a Subspace Trajectory for an Arbitrary Operator

If the operator A has only one eigenvalue, then the limit of the trajectory $A^n(V)$ exists for any subspace V. But, in the general case, the limit of trajectory can does not exist, and the question is: what is the conditions on the subspaces V, whose validity implies the existence of the limit of the trajectory. In this chapter, we discuss this problem for an arbitrary linear invertible operator A in a m-dimensional complex space X.

The algebraic approach allows to obtain this answer in principle: the existence of the limit of the trajectory of the subspace V is equivalent to the existence of the limit of trajectory of the one-dimensional subspace L from $\overset{d}{\bigwedge} X$, spanned by a d-vector $\tilde{v}$, under the action of the operator $\overset{d}{\bigwedge} A$; the conditions for the existence of the limit of a trajectory of a one-dimensional subspace L have also been obtained. These conditions consist in the vanishing of some coordinates of the vector $\tilde{v}$ in the Jordan basis for the operator $\overset{d}{\bigwedge} A$. As noted previously, in the realization of this approach, one has to construct the Jordan form for $\overset{d}{\bigwedge} A$, being a sufficiently complex problem. In addition, the geometric meaning of the obtained conditions remains unclear.

In this section we consider the problem from the geometric point of view, and the basic result are conditions for the existence of the limit of the trajectory, formulated in geometric terms.

12.1 Notations and Preliminary Information

We assume, as before, that in a m-dimensional space X there is the Jordan basis for the operator A, for vectors from this basis we use previously introduced special numeration by four indices. We assume that a scalar product is given and the Jordan basis is orthonormal.

© Springer International Publishing Switzerland 2014

Ć.B. Dolićanin and A.B. Antonevich, *Dynamical Systems Generated by Linear Maps*,
DOI: 10.1007/978-3-319-08228-8_12

Note that by $L(k)$ we denote the subspace, spanned by basis vectors $e(k, j, i, l)$ with the given k; by $L(k, j)$ we denote the subspace, spanned by vectors $e(k, j, i, l)$ for given k and j. We denote projectors onto introduced subspaces by $P(k)$ and $P(k, j)$.

There is a decomposition

$$X = \bigoplus_k L(k) = \bigoplus_{k=1}^{q} \bigoplus_{j=1}^{q(k)} L(k, j). \qquad (12.1.1)$$

$$A = \bigoplus_k A(k) = \bigoplus_{k=1}^{q} \bigoplus_{j=1}^{q(k)} A(k, j),$$

where $A(k)$ and $A(k, j)$ are restrictions of the operator A on the corresponding invariant subspaces.

The behavior of trajectories of a d-dimensional subspaces V depends on the position of V with respect to pyramids (2.2.3) and (2.2.8), induced by the given operator. Recall, that the pyramid (2.2.3) consists of subspace

$$M^k = \bigoplus_{j=1}^{k} L(j), \ 1 \leqslant k \leqslant q.$$

One may construct decomposition of the subspace V, related to this pyramid:

$$V = \bigoplus_{k=1}^{q} W(k), \qquad (12.1.2)$$

where $W(k)$ is a subspace of M^k, satisfying the condition

$$W(k) \bigcap M^{k-1} = 0.$$

Denote by $K(V)$ the set of values k for wich $W(k) \neq \{0\}$. Further, one may assume that $k \in K(V)$ in all sums with respect to index k, since other summands are 0.

By denoting by $Q(k)$ the projector in V onto the subspace $W(k)$, we obtain a representation of an arbitrary vector from V:

$$x = \sum_{k=1}^{q} w(k), \ \text{where} \ w(k) = Q(k)x \in W(k).$$

If $W(k)$ is a subspace from the decomposition (12.1.2), then the following properties hold obviously for the projection $P(i)W(k)$.

(i) $P(i)W(k) = 0$ for $i > k$,

(ii) $P(k)W(k) = P(k)V$,

(iii) $\dim P(k)W(k) = \dim W(k) = \dim P(k)V$.

Note that in the general case, the subspace $W(k)$ cannot be chosen, as the one belongs to the subspace $L(k)$, therefore the projections $P(i)W(k)$ may be non-zero for $i < k$. This implies that, in the general case,

$$W(k) \neq \bigoplus_{i \leq k} P(i)W(k),$$

although for each $x \in W(k)$, one has

$$x = \sum_{i \leq k} P(i)x.$$

This is related to the fact that projections $P(i)x$ are dependent and this dependence is described by the following Lemma.

Lemma 12.1.1 *For a vector $x \in W(k)$, the projection $P(i)x$ for $i < k$ is uniquely determined by the projection $P(k)x$, and there is such a constant C, hence*

$$\|P(i)x\| \leq C\|P(k)x\|$$

for all $x \in W(k)$ and for all values i and k.

Proof Since $M^{k-1} \subset \ker P(k)$ and

$$W(k) \bigcap M^{k-1} = 0,$$

the projector $P(k)$ bijectively maps subspace $W(k)$ onto the image $P(k)W(k)$, the inverse operator $R(k) : P(k)W(k) \to W(k)$ is defined and this operator is bounded. Therefore, $P(i)x = P(i)R(k)P(k)x$, where

$$\|P(i)x\| \leq \|[P(i)R(k)]\| \|P(k)x\|.$$

By putting $C = \max_{i,k} \|[P(i)R(k)]\|$, we obtain the required inequality. □

Further, we also use projections $P(i, j)(W(k)) \subset L(i, j)$. Note that, similarly, one has the inclusion

$$W(k) \subset \bigoplus_{ij} P(i, j)(W(k)),$$

and the equality does need not to be hold.

Note that, on the Grassmann manifold, we consider the metric

$$\varrho(V, W) = \max_{y \in V, \|y\|=1} \min_{x \in W} \|x - y\| = \max_{x \in W, \|x\|=1} \min_{y \in V} \|x - y\| = \|P_V - P_W\|,$$

where P_V is the orthogonal projector onto the subspace V.

Lemma 12.1.2 *Let U be a unitary operator in the space X. Then the action of the induced map φ_U on the Grassmann manifold is isometric, i.e.*

$$\varrho(U(V), U(W)) = \varrho(V, W).$$

Proof If P_V is the orthogonal projector onto the subspace V, then the operator $P_{U(V)} = U P_V U^{-1}$ is the orthogonal projector onto the subspace $U(V)$. Therefore,

$$\varrho(U(V), U(W)) = \|P_{U(V)} - P_{U(W)}\| = \|U(P_V - P_W)U^{-1}\|$$
$$= \|P_V - P_W\| = \varrho(V, W). \qquad \square$$

12.2 The Convergence of the Projections of a Sequence of Subspaces

We are studying the relation between the convergence of a trajectory of a subspace with the convergence of its projections. Firstly, we point out that in general case, the existence of the limit of a sequence of subspaces is not related to the existence of limit of subspaces formed by their projections. Let us show this, in examples.

Suppose that in a subspace X one has a pair of projectors P and Q, where $P + Q = I$, and one has a sequence of subspaces V_n. It seems plausible that the following equation holds

$$\lim_{n \to \infty} (V_n) = \lim_{n \to \infty} P(V_n) \bigoplus \lim_{n \to \infty} Q(V_n). \tag{12.2.3}$$

However, in the general case this is not true and different variations are possible: if two of these limits exist, which appear in (12.2.3), it may happen that the third one does not exist; it can also occur that all limits exist, but the equality does not hold. Actually, this property is another formulation of the previously mentioned discontinuity of the map (10.1.1).

Example 12.2.1 Consider two-dimensional subspace X with orthogonal basis vectors e_1 and e_2. Let P be the orthogonal projector onto the subspace L_1, spanned by the vector e_1, Q—the orthogonal projector onto the subspace L_2, spanned by the vector e_2.

Consider a sequence V_n of one-dimensional subspaces, where V_n is spanned by the vector

$$e_1 + \frac{1}{n}e_2.$$

Then the sequence V_n converges to the subspace $V_0 = L_1$, spanned by the vector e_1. In this example we have $P(V_n) = L_1$ and

$$\lim_{n\to\infty} P(V_n) = P(V_0) = L_1.$$

But, $Q(V_n) = L_2$ for any n, $Q(V_0) = \{0\}$ and

$$\lim_{n\to\infty} Q(V_n) = L_2 \neq Q(V_0) = \{0\}.$$

In this way, the equality (12.2.3) does not hold, although all three limits do exist.

Example 12.2.2 Let us consider two-dimensional subspace and projectors, as in Example 12.2.1. Let V_n be a sequence of one-dimensional subspaces, where the subspace V_n is spanned by the vector

$$\sin n e_1 + \cos n e_2.$$

In this example, we have $P(V_n) = L_1$, $Q(V_n) = L_2$ and the limits of projections do exist:

$$\lim_{n\to\infty} P(V_n) = L_1, \ \lim_{n\to\infty} Q(V_n) = L_2.$$

However, the sequence of subspaces V_n does not have a limit (subspaces V_n "rotates").

This example shows that from the existence of the limit of a sequence of projections does not follow the existence of the limit of a sequence of subspaces.

Example 12.2.3 Consider a three-dimensional space X with orthonormal basis vectors e_1, e_2, e_3. Let P be an orthogonal projector onto the subspace L_1, spanned by the vector e_1, Q—the orthogonal projector onto two-dimensional subspace L_{23}, spanned by the vectors e_2 and e_3.

Consider the sequence V_n of one-dimensional subspaces, where V_n is spanned by the vector

$$e_1 + \frac{1}{n}\sin n \, e_2 + \frac{1}{n}\cos n \, e_3.$$

Then, the sequence V_n converges to the subspace $V_0 = L_1$, spanned by the vector e_1. In this example, we have $P(V_n) = L_1$ and

$$\lim_{n\to\infty} P(V_n) = P(V_0) = L_1.$$

However, $Q(V_n)$ is a subspace, spanned by the vector

$$\sin ne_2 + \cos ne_3,$$

and the sequence of projections $Q(V_n)$ does not have a limit (as in Example 12.2.2).

This example shows that from the existence of the limit of sequence of subspaces, and the existence of limit of first projections, does not follow the existence of the limit of second the projections.

In this way, in general case, the behavior of projections does not allow to make determined conclusions about the behavior of the sequence of subspaces, and theorems on the relation between the convergence of projections and the convergence of sequence of original subspaces may only be obtained under some additional conditions.

Lemma 12.2.4 *Suppose that X is decomposed into an orthogonal direct sum of subspaces L_k, P_k—corresponding projectors, and let subspaces V and W are given, decomposing into a direct sum of their projections:*

$$V = \bigoplus_k P_k(V), \quad W = \bigoplus_k P_k(W).$$

Then,

$$\varrho(V, W) = \max_k \varrho(P_k(V), P_k(W)). \tag{12.2.4}$$

Proof The projector onto the subspace $P_k(V)$ is the operator $P_k P_V$, the projector onto the subspace $P_k(W)$ is the operator $P_k P_W$. Therefore,

$$P_V - P_W = \bigoplus_k (P_k P_V - P_k P_W),$$

and

$$\varrho(V, W) = \| P_V - P_W \| = \max_k \| P_k P_V - P_k P_W \| = \max_k d(P_k(V), P_k(W)).$$

$\square$

Note that the obtained equality may be seen as a strengthen triangle inequality for orthogonal sums of subspaces:

$$\varrho\left(\bigoplus_k V_k, \bigoplus_k W_k\right) \le \max_k \varrho(V_k, W_k).$$

Lemma 12.2.5 *Suppose that the space X is decomposed into an orthogonal direct sum of subspaces, $L(k)$, $P(k)$—the corresponding projectors and let be given a* *sequence V_n of subspaces of X, such that each subspace V_n is decomposed into a direct sum of its projections:*

$$V_n = \bigoplus_k P(k)(V_n).$$

The sequence V_n has the limit $\tilde{V}$ if and only if for each k there exists

$$\lim_{n \to \infty} P(k)(V_n) := \tilde{V}_k,$$

and then

$$\tilde{V} = \bigoplus_k \tilde{V}_k.$$

Proof According to (12.2.4), we have equality

$$d(V_n, V_s) = \max_k d(P_k(V_n), P_k(V_s)),$$

where the claim of the Lemma follows. □

12.3 Reduction to the Convergence of Projections $P(k)(W(k))$

Let us consider decomposition (12.1.2) of the subspace V:

$$V = \bigoplus_{k=1}^{q} W(k),$$

where $W(k)$ is a subspace of M^k, satisfying the condition

$$W(k) \bigcap M^{k-1} = 0.$$

Here the spaces does not orthogonal and we can not applay the Lemma 12.2.5.

Lemma 12.3.1 *Let $W(k)$ be a subspace of M^k such that*

$$W(k) \bigcap M^{k-1} = 0.$$

The trajectory $A^n(W(k))$ of the subspace $W(k)$ approaches the trajectory $A^n(P(k)(W(k)))$ of projection $P(k)(W(k))$.

Proof Let us first note that for the given operator A there are constants C_1 and p such that for all $x \in M(k)$ one has the estimate

$$r(k)^n \|x\| \le \|A(k)^n x\| \le C_1 r(k)^n n^p \|x\|.$$

It follows from the decomposition of vector trajectory, obtained before. Here as p one may take the largest of dimensions of the Jordan cells.

For $x \in W(k)$, we have the decomposition

$$x = w + \sum_{i<k} P(i)x,$$

where

$$w = P(k)x \in P(k)(W(k)).$$

Therefore,

$$A^n x = A(k)^n w + \sum_{i<k} A(i)^n P(i)x.$$

As noted before, the projector $P(k)$ bijectively maps $W(k)$ onto $P(k)(W(k))$. Therefore, there is a constant C_2 where

$$C_2 \|x\| \le \|P(k)x\|, \quad x \in W(k).$$

Note also, that one has the inequality

$$\|P(i)x\| \le C \|P(k)x\|.$$

Then,

$$\|A^n(k)P(k)x\| \ge r(k)^n \|P(k)x\| \ge C_2 r(k)^n \|x\|,$$

and for $i < k$ we have

$$\|A(i)^n P(i)x\| \le C_1 r(k-1)^n n^p \|P(i)x\| \le C_1 C r(k-1)^n n^p \|P(k)x\|.$$

Therefore,

$$\|\frac{1}{\|A^n w\|} A^n x - \frac{1}{\|A^n w\|} A^n w\| = \frac{1}{\|A^n w\|} \|\sum_{i<k} A(i)^n P(i)x\| \tag{12.3.5}$$
$$\le \frac{C_1 C}{C_2} [\frac{r(k-1)}{r(k)}]^n n^p.$$

An arbitrary vector of unit norm from the subspace $A^n(P(k)(W(k)))$ may be represented in the form $\|A^n w\|^{-1} A^n w$, the vector $\|A^n w\|^{-1} A^n x$ belongs to the subspace $A^n(W(k))$. So, from the inequality (12.3.5), and the definition of the metric, it follows that

$$\varrho(A^n(W(k)), A^n(P(k)W(k)) \to 0. \qquad \square$$

Theorem 12.3.2 *Under the decomposition (12.1.2), the trajectory $A^n(V)$ of the subspace V has a limit $\widetilde{V}$ if and only if for each k there exists a limit $\widetilde{W}_k$ of the trajectory of the subspace $P(k)(W(k))$, and then*

$$\widetilde{V} = \lim_{n \to +\infty} A^n(V) = \bigoplus_k \widetilde{W}_k.$$

Proof We consider the trajectory $\widehat{V}_n = A^n(\widehat{V})$ of the subspace

$$\widehat{V} = \bigoplus_{k=1}^{q} P(k)(W(k)).$$

Since the projectors $P(k)$ commute with operator A, we obtain that each subspace $\widehat{V}_n$ decomposes into a direct sum, by the same system of projectors $P(k)$:

$$\widehat{V}_n = \bigoplus_{k=1}^{q} A^n[P(k)(W(k))] = \bigoplus_{k=1}^{q} P(k)A^n[P(k)(W(k))].$$

Therefore, one may apply Lemma 12.2.5, and it follows that the limit of the trajectory $A^n(\widehat{V})$ exists if and only if for each k there exists the limit of a trajectory of subspaces $A^n[P(k)W(k)]$, and then

$$\lim_{n \to \infty} A^n(\widehat{V}) = \bigoplus_{k=1}^{q} \lim_{n \to \infty} A^n[P(k)W(k)] = \bigoplus_k \widetilde{W}_k. \qquad (12.3.6)$$

According to Lemma 12.3.1, the limit of trajectory $A^n(W(k))$ coincides with $\widetilde{W}_k$—the limit of the trajectory of the subspace $P(k)(W(k))$.

Therefore, if there is the limit (12.3.6), then there is also

$$\lim_{n \to \infty} A^n(V) = \bigoplus_k \widetilde{W}_k.$$

Suppose that

$$\lim_{n \to \infty} A^n(V)$$

exists. Subspaces $A^n(W(k))$ are not pairwise orthogonal, so we cannot apply Lemma 12.2.5 to the decomposition (12.1.2). But, according to Lemma 12.3.1, for large n subspace $A^n(W(k))$ is close to the subspace $A^n(P(k)(W(k)))$, and these subspaces are pairwise orthogonal. Therefore, for large n subspaces $A^n(W(k))$ are almost orthogonal and, similarly to the Lemma 12.2.5, from the convergence of $A^n(V)$ convergence of components $A^n(W(k))$ follows. $\qquad \square$

12.4 The Case of Eigenvalues of Equal Moduli

According to the Theorem 12.3.2, the problem on the convergence of the trajectory
of subspace is reduced to the problem on the convergence of trajectories of subspaces
$P(k)(W(k))$. Subspace $P(k)(W(k))$ belongs to the subspace $L(k)$, where the operator $A(k)$ acts, and for this operator moduli of all eigenvalues are equal. Consider
trajectories of subspaces under an action of such an operator.

Suppose that all eigenvalues of the operator A have the moduli equal to r. At the
same time, the operator may have different eigenvalues and it need not be diagonalizable, i.e. it may have non-trivial Jordan cells. This situation is more complex
than the one from the previously discussed case, when all eigenvalues were equal.
In particular, there are subspaces whose trajectories do not have a limit.

Under the given assumptions, the index k, the indexing moduli of eigenvalues in
the special numeration of the Jordan basis, has only one value 1. In the introduced
notation, eigenvalues are

$$\lambda(1, 1), \lambda(1, 2), \ldots, \lambda(1, q(1)).$$

Decomposition of the space X turns into a decomposition of the form

$$X = \bigoplus_{j=1}^{q(1)} L(1, j).$$

The operator A decomposes into a direct sum

$$A = \bigoplus_{j=1}^{q(1)} A(1, j),$$

where each operator $A(1, j)$ has only one eigenvalue $\lambda(1, j)$, but it may have more
than one Jordan cells.

Let $P(1, j)$ be the corresponding projector onto the subspace $L(1, j)$. We define
a unitary operator U, acting by the formula

$$Ux = \sum_{j=1}^{q(1)} \omega_j P(1, j)x, \quad \text{where} \quad \omega_j = \frac{\lambda(1, j)}{r}.$$

We actually consider the previously obtained representation

$$A(k) = rU(k) + N(k),$$

where operator $U(k)$ is unitary and operator $N(k)$ nilpotent.

Consider the operator

$$B = U^{-1}A.$$

All eigenvalues of operator B are equal to r. According to the Theorem 9.1.4, the trajectory of an arbitrary subspace V under the action of the operator B has limit $\widetilde{V}$, and this limit may be constructed using linear algebra operations.

Theorem 12.4.1 *Suppose that all moduli of eigenvalues of operator A are equal to r, B and U are previously introduced operators, and*

$$\widetilde{V} = \lim_{n \to \infty} B^n(V).$$

The trajectory $A^n(V)$ has the limit if and only if the subspace $\widetilde{V}$ satisfies one of three equivalent conditions:

(i) $\widetilde{V}$ is invariant with respect to U;
(ii) $\widetilde{V}$ is invariant with respect to A;
(iii)

$$\widetilde{V} = \bigoplus_j P(1, j)(\widetilde{V}).$$

If these conditions are satisfied, then $\lim A^n(V) = \widetilde{V}$.

Proof By the construction we have $A = UB$, where the factors commute. Therefore, $A^n = U^n B^n$ and, in particular,

$$A^n(V) = U^n B^n(V).$$

For the sequence of subspaces $B^n(V)$, there is the limit $\widetilde{V}$:

$$\varrho(B^n(V), \widetilde{V}) \to 0.$$

The operator U is unitary, so, by the Lemma 12.1.2,

$$\varrho(A^n(V), U^n(\widetilde{V})) = \varrho(U^n B^n(V), U^n(\widetilde{V})) = \varrho(B^n(V), \widetilde{V}) \to 0.$$

This means that the sequence of subspaces $A^n(V)$ approaches to the sequence of subspaces $U^n(\widetilde{V})$. Therefore, from the existence of a limit for one of these sequences, it follows the existence of a limit for the other one.

The sequence $U^n(\widetilde{V})$ is a trajectory of the subspace $\widetilde{V}$ under the action of a unitary operator U. Under the action of a unitary operator, trajectory $U^n(\widetilde{V})$ has a limit if and only if it is stationary, i.e. when this subspace is invariant with respect to the action of U.

In this way, the limit of the trajectory $A^n(V)$ exists if and only if the subspace $\widetilde{V}$ is invariant with respect to the action of B. Therefore, the invariance of $\widetilde{V}$ with respect to the action of U is equivalent to the invariance with respect to the action of the operator A.

On the subspace $L(1, j)$, the operator U acts as a multiplication by a constant and each subspace of $L(1, j)$ is invariant with respect to U. Therefore, the equivalence of conditions (i) and (iii) follows from the Lemma 6.2.1.   □

The condition for the existence of the limit in this Theorem uses subspace $\tilde{V}$—the limit of the trajectory of a subspace V under the action of an auxiliary operator B. The space $\tilde{V}$ depends on V in a complicated way, and in general case we cannot obtain conditions for the existence of the limit as a condition on V in an explicit form. However, one may give simple sufficient conditions on V, where there is the limit of a trajectory.

Corollary 12.4.2 *If*

$$V = \bigoplus_j P(1, j)(V), \tag{12.4.7}$$

then

$$\lim_{n \to \infty} A^n(V) = \bigoplus_{j=1}^{q(1)} \lim_{n \to \infty} A(1, j)^n[P(1, j)(V)] = \tilde{V},$$

where all considered limits exist.

Let us illustrate this Theorem in an example which shows, in particular, that the condition (12.4.7) of decomposability in a direct sum is not necessary for the existence of the limit of a trajectory of a subspace. If the subspace V is not decomposable into a direct sum, then there are relations among the projections $P(1, j)x$. In this way, it follows from the Theorem that the limit of the trajectory of a subspace V exists if all relations among the projections $P(1, j)(A^n x)$ vanish in the limit.

Example 12.4.3 As an example, let us consider an operator in the space $X = \mathbb{C}^4$, given by the matrix

$$A = \begin{pmatrix} 1 & 1 & 0 & 0 \\ 0 & 1 & 0 & 0 \\ 0 & 0 & -1 & 1 \\ 0 & 0 & 0 & -1 \end{pmatrix},$$

consisting of two Jordan cells, with eigenvalues 1 and -1.

In this case, the operator B is given by the matrix

$$B = \begin{pmatrix} 1 & 1 & 0 & 0 \\ 0 & 1 & 0 & 0 \\ 0 & 0 & 1 & 1 \\ 0 & 0 & 0 & 1 \end{pmatrix},$$

i.e. it is the model operator (6.4.8).

For the operator A the set $Inv_2(B)$ of two-dimensional invariant subspaces contains only three subspaces: V_{12}, V_{34} and V_{13}, spanned by the corresponding basis vectors.

For the operator B the set $Inv_2(B)$ of two-dimensional invariant subspaces is extensive, it is parametrized by two complex parameters. At the same time, as shown before, for any $V \in (G(4, 2) \setminus Inv_2(B))$, the limit of the trajectory $B^n(V)$ is the subspace V_{13}. Since the subspace V_{13} is invariant with respect to A, for any $V \in (G(4, 2) \setminus Inv(B))$ there is the limit of the trajectory $A^n(V)$, and this limit is the subspace V_{13}.

The limit of the trajectory $A^n(V)$ also exists for invariant subspaces V_{12} and V_{34}. And for other subspaces, invariant with respect to B, the limit of the trajectory $A^n(V)$ does not exist.

Let us consider, in more details, the two-dimensional subspace V, spanned by the vectors $v_1 = (1, 0, 0, 1)$ and $v_2 = (0, 1, 1, 0)$, i.e. the subspace consisting of vectors $x = (x_1, x_2, x_3, x_4)$ of the form $x = (\xi_1, \xi_2, \xi_2, \xi_1)$. The limit of the trajectory of the subspace V, under the action of operator B, is the subspace V_{13}. This subspace is invariant for A, and it is a direct sum of its projections.

At the same time, the subspace V is not a direct sum of its projections. For $x \in V$, we have $y_1 = P(1, 1)x = (\xi_1, \xi_2, 0, 0)$, $y_2 = P(1, 2)x = (0, 0, \xi_2, \xi_1)$ and there is a dependence among these projections: if $y_1 = (y_1(1), y_1(2), y_1(3), y_1(4))$, $y_2 = (y_2(1), y_2(2), y_2(3), y_2(4))$, then $y_1(1) = y_2(4)$, $y_1(2) = y_2(3)$.

The similar relation exists among the projections of the vectors $\eta = (\eta_1, \eta_2, \eta_3, \eta_4)$ from $A^n(V)$. Such a vector is represented in the form

$$\eta = A^n x = (\xi_1 + n\xi_2, \xi_2, (-1)^n[\xi_2 - n\xi_1], (-1)^n\xi_1),$$

its projections are

$$P(1, 1)\eta = (\eta_1, \eta_2, 0, 0) = (\xi_1 + n\xi_2, \xi_2, (-1)^n[\xi_2 - n\xi_1], (-1)^n\xi_1),$$

$$P(1, 2)\eta = (0, 0, \eta_3, \eta_4) = (\xi_1 + n\xi_2, \xi_2, (-1)^n[\xi_2 - n\xi_1], (-1)^n\xi_1).$$

Here the relation among the projections are

$$\frac{1}{n}\eta_1 - \eta_2 = \frac{(-1)^n}{n}\eta_4,$$

$$\frac{1}{n}\eta_3 + \eta_4 = \frac{(-1)^n}{n}\eta_2.$$

In the limit, the first of these relations transforms into an equality $\eta_2 = 0$, which is not a relation among projections, but it is a condition for the first projection of $(\eta_1, \eta_2, 0, 0)$. Similarly, the second relation in the limit transforms into an equality $\eta_4 = 0$.

12.5 The Main Theorem

Summarizing the above results, we obtain a solution of the problem on the existence of limits of trajectories of subspaces for an arbitrary operator. We use previous notation. In particular, the operator A is representable in the form of a direct sum

$$A = \bigoplus_k A(k) = \bigoplus_{k=1}^{q} \bigoplus_{j=1}^{q(k)} A(k, j),$$

where $A(k)$ and $A(k, j)$ are restriction of A onto corresponding invariant subspaces. To this operator, we associate the operator

$$B = \bigoplus_{k=1}^{q} \bigoplus_{j=1}^{q(k)} \frac{r(k)}{\lambda(k, j)} A(k, j) = \bigoplus_{k=1}^{q} B(k).$$

Theorem 12.5.1 *Let*

$$V = \bigoplus_{k=1}^{q} W(k)$$

be a decomposition of the subspace V, induced by the chain of subspaces $L(k)$, and let

$$\widehat{V} = \bigoplus_{k=1}^{q} P(k)(W(k)).$$

A limit $\widetilde{V}$ of the trajectory $B^n(\widehat{V})$ exists and

$$\widetilde{V} = \lim_{n \to \infty} B^n(V) = \lim_{n \to \infty} B^n(\widehat{V}) = \bigoplus_{k=1}^{q} \widetilde{W}_k,$$

where

$$\widetilde{W}_k = \lim_{n \to \infty} B(k)^n [P(k)W(k)]$$

and these subspaces may be constructed by linear algebra operations.
    The limit of the trajectory $A^n(V)$ exists if and only if for each k the subspace $\widetilde{W}_k$ is decomposable into a direct sum of its projections

$$\widetilde{W}_k = \bigoplus_{j=1}^{q(k)} P(k, j)(\widetilde{W}_k), \qquad (12.5.8)$$

and then

$$\lim_{n\to\infty} A^n(V) = \lim_{n\to\infty} B^n(V) = \widetilde{W}.$$

Let us list a few simple sufficient conditions for the existence of the limit.

Corollary 12.5.2 *Suppose that all moduli of eigenvalues for the operator A are different. Then, for any subspace V, there is the limit of the trajectory and*

$$\lim_{n\to\infty} A^n(V) = \bigoplus_{k} [\lim_{n\to\infty} A^n[P(k)W(k)].$$

Proof According to the condition, by the previously introduced notation, the index j has only value 1. Therefore, for any k in the decomposition of the subspace $\widetilde{W}_k$, there is only one summand and the equality (12.5.8) holds. □

We have formulated this statement without proof as the Theorem 9.1.5. As it has been explained, this result is interesting as an example of the claim which cannot be obtained directly by the use of the algebraic approach.

Corollary 12.5.3 *Let A be an arbitrary invertible linear operator. If for each k in the decomposition (12.1.2) the equality*

$$P(k)W(k) = \bigoplus_{j} P(k, j)W(k)$$

holds, then

$$\lim_{n\to\infty} A^n(V) = \bigoplus_{k}\bigoplus_{j} \lim_{n\to\infty} A^n[P(k, j)W(k)]$$

and all written limits exist.

Proof The trajectory of the subspace V approaches toward the trajectory of the subspace $\widehat{V}$, which, in the discussed case, is represented in the form of a direct sum

$$\widehat{V} = \bigoplus_{k}\bigoplus_{j} P(k, j)(W(k)).$$

We have analogous decomposition as a direct sum for subspaces $A^n(\widehat{V})$. For any of subspaces $P(k, j)(W(k))$ there is the limit of the trajectory. Therefore, one may apply the Lemma 12.2.5 with the results follow. □

Part III
The Applications to Weighted Shift Operators

Chapter 13
The Spectrum of Weighted Shift Operator

As an application of the results on the dynamics of a linear map some results on spectral properties of weighted shift operators are presented below.

13.1 The Weighted Shift Operators

A bounded linear operator B, acting on the Banach space $F(X)$ of functions or vector-valued functions on an arbitrary set X is called a *weighted shift operator* (WSO), if it can be represented in the form

$$Bu(x) = a_0(x)u(\alpha(x)), \quad x \in X, \tag{13.1.1}$$

where $\alpha : X \to X$ is a map, $a_0(x)$ a scalar or matrix-valued function on X.

Such operators, operator algebras generated by them and functional equations related to such operators have been studied by many authors in various functional spaces, as an independent object and in relation to various applications (theory of dynamical systems, integro-functional, differential-functional, functional and difference equations, automorphisms and endomorphisms of the Banach algebras, nonlocal boundary value problems, nonclassical boundary value problems for equation of string vibration, the general theory of operator algebras, etc. (see [1–13]).

The properties of WSO depend on all the data: the space of functions under investigation, the form of coefficients and on the map α. The main problem, in this subject, is related to the establishment of relations between the spectral properties of weighted shift operators and the behavior of trajectories, i.e. dynamic properties of the map α.

We note first that in the study of WSO, the special and rather simplest case is when the map α is periodic. But under the periodicity condition, there are a lot of delicate questions too (see, for example, [8]).

The problem of describing the spectrum of a weighted shift operator, in the case of invertible mapping α in classical spaces, is fundamentally solved in sufficient generality. In these spaces, the operator B can be conveniently written in the form

© Springer International Publishing Switzerland 2014

Ć.B. Dolićanin and A.B. Antonevich, *Dynamical Systems Generated by Linear Maps*,
DOI: 10.1007/978-3-319-08228-8_13

of $B = aT_\alpha$, introducing an auxiliary operator of the form $T_\alpha u(x) = \varrho(x)u(\alpha(x))$, where the *normalizing function* ϱ is selected, depending on the space $F(X)$, such a way, that the operator T_α is isometric.

We restrict ourselves below to operators in the space $L_2(X, \mu)$, so we formulate the results known only for this case, without dealing with their generalizations to other classes of spaces.

For spaces $L_2(X, \mu)$, the normalizing function ϱ exists if and only if the map α preserves the class of the measure μ (i.e. if for a measurable set E equality $\mu(\alpha^{-1}(E)) = 0$ holds if and only if $\mu(E) = 0$). If X is $\mathbb{R}^n$, the map α is a diffeomorphism and $F(X) = L_2(X)$, then $\varrho(x) = |J_\alpha(x)|^{1/2}$, where $J(\alpha)$ is the Jacobian of the diffeomorphism α.

When writing the operator in the form of $B = aT_\alpha$, the properties of the operator B are more simply expressed using the *reduced coefficient*

$$a(x) = \varrho(x)^{-1}a_0(x).$$

The description of the spectrum in the space of scalar functions uses the following notions.

A measure ν on X is called $\alpha-invariant$, if

$$\nu(\alpha^{-1}(\omega)) = \nu(\omega)$$

for all measurable ω.

A measure ν on X is called *ergodic with respect to* α, if from equality $\alpha^{-1}(\omega) = \omega$ follows either $\nu(\omega) = 0$ or $\nu(X \setminus \omega) = 0$.

For a given map $\alpha : X \to X$, a topological space X is called α-*connected* if cannot be decomposed into two nonempty closed subsets invariant with respect to α. Each connected space is α-connected, but α-connected space may be disconnected.

The general result on the spectrum of weighted shift operator is given by the next theorem presented in [1, 3].

Theorem 13.1.1 *Let X be a compact space, μ—a measure on X, whose support coincides with the whole space, $\alpha : X \to X$—invertible continuous map, preserving the class of measure μ, $a \in C(X)$. In the space $L_2(X, \mu)$, for the spectral radius $R(B)$ for the operator $B = aT_\alpha$ we have*

$$R(B) = \max_{\nu \in M_\alpha(X)} \exp \left[\int_X \ln |a(x)| d\nu \right], \tag{13.1.2}$$

where a is the reduced coefficient, and $M_\alpha(X)$ is the set of probability measures on X, invariant and ergodic with respect to the map α.

If $a(x) \neq 0$ for all x, then the spectrum $\sigma(B)$ belongs to the ring

$$K = \{\lambda \in \mathbb{C} : r(B) \leq |\lambda| \leq R(B)\}, \tag{13.1.3}$$

where

$$r(B) = \min_{\nu \in M_\alpha(X)} \exp\left[\int_X \ln |a(x)| d\nu\right],$$

and K is the smallest ring with the center at 0, containing the spectrum.

If the set of nonperiodic points of the map α is everywhere dense in X, and the space X is α-connected, then the spectrum coincides with the ring K, and in the case when $a(x) = 0$ at least at one point, the spectrum $\sigma(B)$ is the circle of radius $R(B)$ with the center at the point 0.

Note that this Theorem has a conditional character—it reduces the description of the spectrum to the construction of the set M_α of ergodic measures. Such a set depends on the dynamic of the map, and to find it for an arbitrary α is rather difficult.

There are a number of specific classes of mappings where an explicit description of the set $M_\alpha(X)$ and, thus, an explicit description of the spectrum of a weighted shift operator are obtained (see, for example [14, 15]). One of these classes form the maps of the Morse-Smale type.

A map α will be called a *map of Morse-Smale type*, if for α the set $Fix(\alpha)$ of periodic points is finite, and the trajectory of each point tends to a trajectory of one of the periodic points as $n \to +\infty$, and tends to a trajectory of a periodic points (possibly different) as $n \to -\infty$.

For such maps, the set $M_\alpha(X)$ consists of finitely many measures, each of which is concentrated on the trajectory of one of the periodic points. In addition, for such maps, the set of nonperiodic points is dense in X (with the exception of degenerate cases, when the fixed points are isolated points of the space X). Therefore, for maps of the Morse-Smale type, Theorem 13.1.1 gives an explicit description of the spectrum.

13.2 The Weighted Shift Operator Induced by Linear Maps

Let μ be the Lebesgue measure on the space $\mathbb{C}^m = \mathbb{R}^{2m}$ and $L_2(\mathbb{C}^m)$ is the space of equivalence classes of measurable functions with finite norm

$$\|u\| = \left(\int_{\mathbb{R}^m} |u(x)|^2 d\mu\right)^{1/2}.$$

Let A be a nonsingular linear map in $\mathbb{C}^m$. In the space $L_2(\mathbb{C}^m)$, let us consider weighted shift operators induced by this map, i.e. the operators determined by the expression

$$Bu(x) = a_0(x)u(Ax), \tag{13.2.4}$$

where a_0 is a given measurable function.

We show how the dynamics of a linear map A allows us to investigate spectral properties of B. Similar results hold for the spaces $L_2(\mathbb{R}^m)$, but for their formulation one needs the description of the dynamics of the linear mapping in $\mathbb{R}^m$. As noted above, such a description can be obtained as a consequence of the above results. We present general results for $\mathbb{C}^m$, but we shall also consider spaces $\mathbb{R}^m$ in an examples, since in this case the results are more illustrative.

For a linear map $\alpha(x) = Ax$ we have $J_\alpha = \det A$, normalizing function is $\varrho(x) = |\det A|^{1/2}$ and this function does not depend on x, the operator T_α acts by the formula

$$T_\alpha u(x) = |\det A|^{1/2} u(Ax),$$

the reduced coefficient is

$$a(x) = \frac{1}{\sqrt{|\det A|}} a_0(x). \tag{13.2.5}$$

Below, we assume that the operator (13.2.4) is written in the form $B = aT_\alpha$.

The weighted shif operators can be considered also in the weighted spaces $L_2(\mathbb{C}^m; v)$, with norm

$$\|u\|_v = \left(\int_{\mathbb{R}^m} |u(x)|^2 v(x) d\mu \right)^{1/2},$$

where v is a given measurable function such that $v(x) > 0$ is almost everywhere. Such operators with $a_0(x) \equiv 1$ was considered, for example, in [16, 17]. We remark, that the case of weighted spaces can be reduced to the case of the space $L_2(\mathbb{C}^m)$, on the basis of the following propositions.

Proposition 13.2.1 *The operator B, given by the formula (13.2.4) is bounded on the space $L_2(\mathbb{C}^m; v)$ if and only if the function*

$$a(x) = \frac{1}{\sqrt{|\det A|}} a_0(x) \sqrt{\frac{v(x)}{v(Ax)}}$$

is essentially bounded. Under this condition

$$\|B\| = \|a\|_\infty.$$

The operator B is similar to the operator B_1 on the space $L_2(\mathbb{C}^m)$ given by the formula

$$B_1 u(x) = a_0(x) \sqrt{\frac{v(x)}{v(Ax)}} u(Ax) = a(x) T_\alpha u(x);$$

consequently, the spectral properties of theses operators coincide.

Proof The map $W : L_2(\mathbb{C}^m; v) \to L_2(\mathbb{C}^m)$, given by the formula

$$Wu(x) = \sqrt{v(x)}u(x),$$

is an isometric isomorphism of given spaces, since $\|Wu\| = \|u\|_v$. The operator $B_1 = WBW^{-1}$, acting on the space $L_2(\mathbb{C}^m)$, is by definition similar to the operator B. By calculating the product of operators, we get:

$$W^{-1}u(x) = \frac{1}{\sqrt{v(x)}}u(x), \quad BW^{-1}u(x) = \frac{a_0(x)}{\sqrt{v(Ax)}}u(Ax),$$

$$B_1u(x) = WBW^{-1}u(x) = a_0(x)\sqrt{\frac{v(x)}{v(Ax)}}u(Ax).$$

$\square$

Thus, it suffices to study operators on $L_2(\mathbb{C}^m)$. Below, we assume, that the coefficient a is a bounded continuous function.

For completeness and comparison of the results, we give a description of the spectrum of the operator B, in the simplest case of a periodic linear map A. The periodicity condition $A^N = I$ is equivalent to the fact that the matrix of the operator reduces to a diagonal form, and that one has on the diagonal numbers

$$\omega_k = e^{i2\pi \frac{m_k}{N}}.$$

We assume the fraction $\frac{m_k}{N}$ is written with the least common denominator.

Theorem 13.2.2 *Let $A^N = I$, where $A^k \neq I$ for $k < N$. The operator $a(x)u$ $(Ax) - \lambda I$ is invertible, if and only if,*

$$\inf | \prod_{k=0}^{N-1} a(A^k x) - \lambda^N| > 0.$$

Corollary 13.2.3 *Suppose,*

$$\Lambda = \{\lambda \in \mathbb{C} : \exists x, \text{ such that} \lambda^N = \prod_{k=0}^{N-1} a(A^k x)\}$$

The spectrum of the operator B is the closure of the set $\Lambda : \sigma(B) = \overline{\Lambda}$.

In the case of non-periodic mapping, the investigation is significantly more complicated. In particular, we cannot directly apply the Theorem 13.1.1, since the space $\mathbb{C}^m$ is not compact.

The study can be reduced to the case of a compact space by using the compactification of the space $\mathbb{C}^m$ (or $\mathbb{R}^m$). Recall that a *compactification of the space* X_0 is a compact space X, such that X_0 is a dense subspace of X. A compactification exists for any locally compact space X_0 and there may be many different compactifications.

One of compactifications of the space $\mathbb{R}^m$ may be constructed as follows. The space $CB(\mathbb{R}^m)$, consisting of bounded continuous functions on $\mathbb{R}^m$, is a C^*-algebra. By the Gelfand-Naimark Theorem, this algebra is isomorphic to the algebra $C(M)$, consisting of continuous functions on the space M of maximal ideals of algebra $CB(\mathbb{R}^m)$. Under this isomorphism, to a function $a \in CB(\mathbb{R}^m)$ a continuous function $\widehat{a}$ on M corresponds, called the *Gelfand transformation* of the function a. The space M is compact, and $\mathbb{R}^m$ may be embedded in M as a dense set.

Let T be a map of $CB(\mathbb{R}^m)$ into itself, acting by the formula $Tu(x) = u(Ax)$. This map is an automorphism of $CB(\mathbb{R}^m)$ as a C^*-algebra. Since under an automorphism a maximal ideal goes to a maximal ideal, the automorphism T induces a homeomorphism α of the space of maximal ideals and on $\mathbb{R}^m$ this map acts by the formula $\alpha(x) = Ax$. The homeomorphism α is called the *prolongation* of the map A, corresponding to the given compactification.

On the compact space M, we give a measure μ so that $\mu(M \backslash \mathbb{R}^m) = 0$, and on $\mathbb{R}^m \subset M$ this measure coincides with the Lebesgue measure. Then $L_2(\mathbb{R}^m) = L_2(M, \mu)$, all the conditions of the Theorem 13.1.1 are satisfied, and one may apply it, to describe the spectrum of the operator.

But, in the practical realization of the formulate plan, some fundamental difficulties arise, since the space M of maximal ideals of algebra $CB(\mathbb{R}^m)$ has a very complex structure, it is not explicitly given and, in addition to that, it is impossible to describe explicitly all invariant measures on that space. Difficulties that emerge are due to the fact that this compactification (called the *Stone-Cech compactification*) is the largest of all the compactifications of the space $\mathbb{R}^m$. Therefore, it is natural to consider a compactification which has a simpler construction, and it is given in an explicit form. Such compactifications can be conducted in an analogous pattern.

Let S be a C^*-subalgebra of $CB(\mathbb{R}^m)$, which separates points of the space $\mathbb{R}^m$, and $M(S)$ be the space of maximal ideals of algebra S. The space $M(S)$ is compact and, according to the Gelfand-Naimark Theorem, the algebra S is isomorphic to the algebra $C(M(S))$, consisting of continuous functions on $M(S)$. Under this isomorphism to a function $a \in S$ a continuous function $\widehat{a}$ on $M(S)$ corresponds, called the Gelfand transformation of the function a. The space $M(S)$ is compact, and $\mathbb{R}^m$ may be embedded as a dense subset of $M(S)$.

With this construction, one can obtain all compactifications of the space $\mathbb{R}^m$. Indeed, let the space $\mathbb{R}^m$ be continuously and densely embedded in a compact space M. To each function $\widehat{a} \in C(M)$, we correspond its restriction to $\mathbb{R}^m$—the function $a(x) = \widehat{a}(x)$, $x \in \mathbb{R}^m$. Then, the set S of these functions a is a C^*-subalgebra of $CB(\mathbb{R}^m)$, separating points of the space $\mathbb{R}^m$, and the space $M(S)$ of maximal ideals of that algebra coincides with M.

However, in the case of an arbitrary compactification of the space $\mathbb{R}^m$, the Theorem 13.1.1 cannot be applied for the study of operators of the form (13.2.4). The additional condition is that the map A may be extended to the homeomorphism of the space X.

This condition is equivalent to the fact that the operator $T_A a(x) = a(Ax)$ is an automorphism of the corresponding algebra S.

Theorem 13.2.4 *The following properties of the topological space M are equivalent:*

(i) *The space M is a compactification of the space $\mathbb{R}^m$, and the map: $x \to Ax, x \in \mathbb{R}^m$, extends to a homeomorphism of the space M;*

(ii) *The space M is a space of maximal ideals of some C^*-subalgebra S of $CB(\mathbb{R}^m)$, separateing points of the space $\mathbb{R}^m$, and S is invariant with respect to actions of operators T_A and T_A^{-1}.*

It follows from this Theorem that the study of the concrete compactification is equivalent to the selection of a subalgebra S containing coefficients of the discussed operators. Such approach permits take into account the complexity of the behavior of the coefficients of weighted shift operators. If the behavior of the coefficients from S is more complex, then the structure of the corresponding space $M(S)$ is more complex, and the more complex is the action of the map α on that space.

Let us discuss two simplest compactifications for which one can obtain a description of the dynamics of the corresponding map α, using the previous results.

13.3 The Compactification by One Point at Infinity

The simplest compactification X of the space $\mathbb{C}^m$ is done by adjoining one point ∞ (Alexandrov's point), whose neighborhoods are given as complements of bounded sets in $\mathbb{R}^m$. The corresponding algebra S consists of continuous functions, having at infinity the limit $a(\infty)$. In this case, the Gelfand transformation acts by the formula

$$\widehat{a}(x) = \begin{cases} a(x), & x \in \mathbb{C}^m, \\ a(\infty), & x = \infty. \end{cases} \tag{13.3.6}$$

This algebra is invariant with respect to any nonsingular linear map A, and the corresponding homeomorphism α acts by the formula

$$\alpha(x) = \begin{cases} Ax, & x \in \mathbb{C}^m, \\ \infty, & x = \infty. \end{cases} \tag{13.3.7}$$

To apply the Theorem, one has to find all probabilities measures on X, invariant and ergodic with respect to the transformation α. One may accomplish this task by using the description of the dynamics of linear map obtained before, and known description of measures on tori, which are invariant and ergodic with respect to the standard shift.

We recall that by $H(\omega)$, we have denoted a closed subgroup of the torus T^d generated by the element $\omega \in T^d$. In any compact group G, there is a unique normalized measure, invariant with respect to translations. This measure is called the *Haar measure*. In the case of the torus ($G = T^d$), the Haar measure is the normalized Lebesgue measure on the torus. In the case under consideration, the group $G = H(\omega)$ is a torus, or a product of a torus and the group $\mathbb{Z}_N$.

Lemma 13.3.1 *Let $\alpha(z) = \omega z$ be the standard shift of the torus T^m, $H(\omega)$ a closed subgroup generated by the element $\omega \in T^m$ and ν_ω be Haar measure on $H(\omega)$. The set of probability measures on T^m, invariant and ergodic with respect to the transformation α is parameterized by elements of the quotient group $T^m/H(\omega)$. Each of such measure is concentrated on one of equivalence classes—an element of the group $T^m/H(\omega)$. This measure on the equivalence class $[z]$, generated by the element z, is an image of the measure ν_ω under the map $H(\omega) \ni \xi \mapsto z\xi \in [z]$.*

In particular, the integral with respect to that measure is given by the formula

$$\int_{[z]} f(x)d\nu = \int_{H(\omega)} f(z\xi)d\nu_\omega.$$

Recall that by Theorem 5.1.1, the set of limit points under action of A consists on point ∞, point 0 and the subspace L_1, generated by all eigenvectors where moduli of eigenvalue is 1. If $r_k \neq 1$ for all k, then $L = \emptyset$. By Theorem 5.2.2, the subspace L splits into tori T_ξ, where action of A is the standard shift. Among these tori, there are those of different dimension, and the standard shift is given by different elements.

Theorem 13.3.2 *Let $A : \mathbb{C}^m \to \mathbb{C}^m$ be a nonsingular linear map and X be the compactification of $\mathbb{C}^m$ by one point at infinity. The set $Mes(X, \alpha)$ contains only the following measures: the measure δ_0, concentrated at the point 0, the measure δ_∞, concentrated at the point ∞ and all measures on tori $T_\xi \subset L_1$, invariant with respect to the corresponding shifts $\omega(\xi)$, described in the Lemma 13.3.1.*

Proof It is known (see [7]), that the support of an ergodic measure coincides with the set of limit points of one of the trajectories. The sets of limit points of the trajectory are described in the Theorems 5.1.1 and 5.2.3. Each of such limit set belongs to some tori $T_\xi \subset L_1$, where the action of the map is of the standard shift. And measures on tori, ergodic and invariant with respect to the standard shift, are described in the Lemma 13.3.1.   □

Thus, the resulting description of the set $Mes(\widetilde{\mathbb{R}}^m, \alpha)$ allows one to find an explicit form the spectrum of corresponding operators. Note that, if the map A is not periodic, then the condition that the set of nonperiodic points is everywhere dense, is satisfied. The condition on the α-connectedness is also satisfied since the space X under consideration is connected.

Here are some typical explicit results for specific classes of A.

1. Suppose that for all eigenvalues of the matrix A, one has $|\lambda(k, j)| \neq 1$. Then, according to the Theorem 5.1.1, this map has two fixed points: $Fix(\alpha) = \{0, \infty\}$, and the trajectory of any vector converges to one of these points. This means that the

map is a map of Morse-Smale type. The set of invariant ergodic measures consists of two measures: $Mes(X, \alpha) = \{\delta_0, \delta_\infty\}$. From the Theorem 13.1.1 we obtain the description of the spectrum of the weighted shift operator.

Theorem 13.3.3 *Let the coefficient a be a continuous function having the finite limit at infinity $a(\infty) \neq 0$ and $a(x) \neq 0$ for all $x \in \mathbb{R}^m$. The spectrum $\sigma(B)$ of the operator B in the space $L_2(\mathbb{C}^m)$ is a ring:*

$$\sigma(B) = \{\lambda \in \mathbb{C} : r(B) \leq |\lambda| \leq R(B)\},$$

where

$$R(B) = \max\{|a(0)|, |a(\infty)|\}, \quad r(B) = \min\{|a(0)|, |a(\infty)|\}.$$

In the given formulation, the dependence of the spectrum in the concrete A is not evident. However, one has to pay attention to the fact that in the example and the examples the following one, we use the reduced coefficient (13.2.5), depending on the determinant of the matrix A. If the operator is given in the form

$$Bu(x) = a_0(x)u(Ax),$$

then, for different A satisfying the condition $|\lambda(k, j)| \neq 1$, we obtain different spectra.

2. Let the matrix A has one simple eigenvalue of modulus 1. Then, this matrix may be reduced to a block-diagonal form

$$A = \begin{pmatrix} e^{i2\pi h} & 0 \\ 0 & A_{m-1} \end{pmatrix}, \tag{13.3.8}$$

where A_{m-1} is a matrix of dimension $m - 1$, where the moduli of eigenvalues are different from 1.

In this example, in addition to measures δ_0 and δ_∞, invariant ergodic measures appear concentrated on circles with the center at the point 0, lying in the plane (one-dimensional complex subspace) L_1, spanned by the first basis vector. The form of these measures (and, consequently, the spectrum of the operator) depends on the number h.

Let us denote by $\psi(a; r)$ the geometric mean of the function $|a|$ on the circle of radius r, lying in the plane L_1:

$$\psi(a; r) = exp\left[\int_0^1 \ln |a(re^{i2\pi t}, 0, \ldots, 0)|dt\right].$$

Theorem 13.3.4 *Let the coefficient a be a continuous function having the finite limit at infinity $a(\infty) \neq 0$ and $a(x) \neq 0$ for all x.*

*1. If the number h is irrational, then the spectrum $\sigma(B)$ of the operator B in the
space $L_2(\mathbb{C}^m)$ is a ring:*

$$\sigma(B) = \{\lambda \in \mathbb{C} : r(B) \le |\lambda| \le R(B)\},$$

where

$$R(B) = \max\{\psi(a; r), 0 < r < \infty; |a(0)|, |a(\infty)|\},$$

$$r(B) = \min\{\psi(a; r), 0 < r < \infty; |a(0)|, |a(\infty)|\}.$$

*2. If the number h is rational and is given in the form of an irreducible fraction
$h = p/N$, then the spectrum $\sigma(B)$ of the operator B in the space $L_2(\mathbb{R}^m)$, is a ring:*

$$\sigma(B) = \{\lambda \in \mathbb{C} : r(B) \le |\lambda| \le R(B)\},$$

where

$$R(B) = \max\{a_h(x_1), x \in \mathbb{C}; |a(0)|, |a(\infty)|\},$$

$$r(B) = \min\{a_h(x_1), x \in \mathbb{C}; |a(0)|, |a(\infty)|\},$$

$$a_h(x_1) = \left[\prod_{k=0}^{N-1} |a(e^{i2\pi kh} x_1, 0, \ldots, 0)|\right]^{1/N}.$$

The quantities $R(B)$ and $r(B)$ completely describe the spectrum and the expression for $r(B)$ in the Theorem 13.1.1 are completely analogous to the expression for $R(B)$ (max is replaced by min). Therefore, in order to shorten statements in the following examples, we only give an expression for $R(B)$.

13.4 The Compactification by the Sphere at Infinity

If one has to study an operator whose coefficient has no limit at infinity, the one has to consider more complex algebras S and, consequently, more complex compactifications of the space $\mathbb{C}^m$ or $\mathbb{R}^m$. We obtain an interesting example by considering compactification by the sphere at infinity.

This compactification arises most clearly by the following construction. The map

$$x \rightarrow \frac{1}{1 + \|x\|} x$$

is a homeomorphism between the space $\mathbb{R}^m$ and the open unit ball. The closed unit ball is a compact space where the open unit ball (homeomorphic to the space $\mathbb{R}^m$) is a dense subset. The subalgebra S corresponding to that compactification is isomorphic to the algebra of functions continuous on the closed unit ball.

Formally, this compactification is constructed as follows. We consider the set

$$X = \widetilde{\mathbb{R}}^m = \mathbb{R}^m \coprod S_\infty^{m-1}$$

and introduce a topology on X. The neighborhood basis of a point $x \in \mathbb{R}^m$ consisting of the balls with the center at that point, the neighborhood basis of a point $\xi_0 \in S_\infty^{m-1}$ consists of the sets

$$W(\xi_0; R, \delta) = \{x \in \mathbb{R}^m : \|x\| > R, \|\frac{1}{\|x\|}x - \xi_0\| < \delta\} \bigcup \{\xi \in S^{m-1}\infty : \|\xi - \xi_0\| < \delta\},$$

where $R > 0$, $\delta > 0$. It is easy to check that this topological space is homeomorphic to the closed unit ball and it is the compactification of the space $\mathbb{R}^m$, obtained by adjoining the sphere at infinity S_∞^{m-1}. In the described compactification the corresponding functional subalgebra S is the algebra of continuous function on $\mathbb{R}^m$ such that for all x, $\|x\| = 1$, there exists the uniform limit

$$\lim_{t \to +\infty} a(t\xi).$$

Then, the function

$$\widehat{a}(x) = \begin{cases} a(x), & x \in \mathbb{R}^m, \\ \lim_{t \to +\infty} a(tx), & x \in S_\infty^{m-1} \end{cases} \tag{13.4.9}$$

is continuous on X.

Lemma 13.4.1 *The algebra S is invariant with respect to a nonsingular linear map. The prolongation α, i.e. a homeomorphism of the space X generated by the linear map A, acts by the formula*

$$\alpha(x) = \begin{cases} Ax, & x \in \mathbb{R}^m, \\ \frac{1}{\|Ax\|}Ax, & x \in S_\infty^{m-1}. \end{cases} \tag{13.4.10}$$

We can now describe properties of weighted shift operators with coefficients from the introduced algebra, since the study of the spectrum of an operator with coefficients from the given algebra S has been reduced to finding measures on X, ergodic with respect to α. Each of such measure is concentrated on the limit set of some trajectory. The limit sets of trajectories on $\mathbb{C}^m$ (and $\mathbb{R}^m$) are described in the Theorem 5.1.1, and the limit sets of trajectories on S_∞^{m-1} are described in the Theorem 5.3.2. In both cases, the limit sets stratify into tori, on which the map α acts as a standard shift. And invariant ergodic measures on tori are described in the Lemma 13.3.1. In this way we obtain a description of the set $Mes(X, \alpha)$, which allows us to find in an explicit form spectra of weighted shift operators with coefficients from algebra S.

The set $Mes(X, \alpha)$ may have different structure depending on properties of operator A, and there are many qualitatively different cases, where the expression for the spectral radius is given by various formulas. For all these cases, the corresponding propositions may be obtained from general theorems and previously obtained description of the dynamics.

For example, even for spaces $\mathbb{C}^2$ and diagonal matrices

$$A = \begin{pmatrix} \lambda_1 & 0 \\ 0 & \lambda_2 \end{pmatrix}$$

with positive eigenvalues $0 < \lambda_1 \leq \lambda_2$, we obtain four different expressions for $R(B)$.

In this case, we consider compactification of the space $\mathbb{C}^2 = \mathbb{R}^4$ by the three-dimensional sphere at infinity. For $a \in S$ we denote the extension of a on S_∞^3 by $\widehat{a}$.

Theorem 13.4.2 *If $a \in S$, $a(x) \neq 0$, $\widehat{a}(x) \neq 0$, then the spectrum $\sigma(B)$ of the operator B in the space $L_2(\mathbb{C}^2)$ is the ring:*

$$\sigma(B) = \{\lambda \in \mathbb{C} : r(B) \leq |\lambda| \leq R(B)\},$$

where numbers $R(B)$ and $r(B)$ are given by different expressions, depending on the relations between λ_1 and λ_2.

1. If $\lambda_1 \neq \lambda_2$, $\lambda_k \neq 1$, then

$$R(B) = \max\{|a(0)|, \max_{|z_1|=1} |\widehat{a}(z_1, 0)|, \max_{|z_2|=1} |\widehat{a}(0, z_2)|\}.$$

2. If $\lambda_1 < \lambda_2 = 1$, the

$$R(B) = \max\{\sup_{x_2 \in \mathbb{C}} |a(0, x_2)|, \max_{|z_1|=1} |\widehat{a}(z_1, 0)|, \max_{|z_2|=1} |\widehat{a}(0, z_2)|\}.$$

3. If $\lambda_1 = \lambda_2 \neq 1$, then

$$R(B) = \max\{|a(0, 0)|, |\widehat{a}(\xi)|, \xi \in S_\infty^3.\}.$$

4. If $1 = \lambda_1 < \lambda_2$, then

$$R(B) = \max\{\sup_{x_1 \in \mathbb{C}} |a(x_1, 0)|, \max_{|z_1|=1} |\widehat{a}(z_1, 0)|, \max_{|z_2|=1} |\widehat{a}(0, z_2)|\}.$$

Remark If $\lambda_1 = \lambda_2 = 1$, then all points are fixed and the condition that the set of nonperiodic points is dense, is not satisfied. In this case, the weighted shift operator degenerates into an operator of multiplication by a function, the spectrum is not invariant with respect to rotation.

As an example, we give here explicit results for some other specific classes of A.

1. Let $A : \mathbb{R}^m \to \mathbb{R}^m$ and let all eigenvalues of the matrix A are positive, different and have multiplicity 1, i.e. in the Jordan form to each eigenvalue corresponds one Jordan cell (*weakened Perron condition* is satisfied). If, in addition to that, all eigenvalues are different from 1, then the map α is a map of the Morse-Smale type, having $2q + 1$ fixed point:

$$Fix(\alpha) = \{0 \in \mathbb{R}^m, \pm e(k, 1) \in S_\infty^{m-1}, k = 1, 2, \ldots, q\}.$$

So, from the Theorem 13.1.1, we obtain a description of the spectrum of the weighted shift operator.

Theorem 13.4.3 *If $a \in S$, $\widehat{a}(x) \neq 0$ for all $x \in X_0$ and the matrix A the weakened Perron condition is satisfied, then the spectrum $\sigma(B)$ of the operator B in the space $L_2(\mathbb{R}^m)$ is the ring*

$$\sigma(B) = \{\lambda \in \mathbb{C} : r(B) \leq |\lambda| \leq R(B)\},$$

where

$$R(B) = \max\{|a(0)|, |\widehat{a}(\pm e(k, 1)|, k = 1, 2, \ldots, q\},$$

$$r(B) = \min\{|a(0)|, |\widehat{a}(\pm e(k, 1))|, k = 1, 2, \ldots, q\}.$$

2. If number 1 is an eigenvalue of the matrix A, then the description of the spectrum changes. For example, the following proposition is true.

Theorem 13.4.4 *Suppose that all eigenvalues of the matrix A are positive, different, have multiplicity 1 and $\lambda_{k_0} = 1$. If $a \in S$, $\widehat{a}(x) \neq 0$ for all x, then the spectrum $\sigma(B)$ of the operator B in the space $L_2(\mathbb{R}^m)$ is the ring*

$$\sigma(B) = \{\lambda \in \mathbb{C} : r(B) \leq |\lambda| \leq R(B)\},$$

where

$$R(B) = \max\{|a(x)|, x \in L_{k_0}; |\widehat{a}(\pm e(k, 1)|, k = 1, 2, \ldots, q\},$$

$$r(B) = \min\{|a(x)|, x \in L_{k_0}; |\widehat{a}(\pm e(k, 1))|, k = 1, 2, \ldots, q\},$$

and L_{k_0} is one-dimensional subspace, generated by corresponding eigenvectors.

4. If the matrix A has complex eigenvalues, then the description of the spectrum of a weighted shift operator changes, the number of qualitatively different cases

increases, and ergodic measures concentrated on tori appear. As an example, let us discuss the operators B, generated by diagonal matrices

$$A = diag(\lambda_1, \lambda_2, \ldots, \lambda_m). \tag{13.4.11}$$

Let us state the result only for one of possible cases. We introduce notation:

$$D_k(a) = \max_{|z_k|=1} |\widehat{a}(0, \ldots, 0, z_k, 0, \ldots, 0)|,$$

$$\psi(a) = exp\left[\int_0^1 \ln |\widehat{a}(e^{i2\pi t}, 0, \ldots, 0)|dt\right],$$

$$\psi(a; s) = exp\left[\int_0^1 \ln |a(se^{i2\pi t}, 0, \ldots, 0)|dt\right].$$

Theorem 13.4.5 *Suppose that* $\lambda_1 = re^{i2\pi h}$, *where h is irrational and all other eigenvalues are positive, different and not equal to 1. The spectral radius* $R(B)$ *of the operator B in the space* $L_2(\mathbb{C}^m)$ *is given by the following expressions.*
1. If $r \neq 1$, *then*

$$R(B) = \max\{|a(0)|, \psi(a), D_2(a), D_3(a), \ldots, D_m(a)\}.$$

2. If $r = 1$, *then*

$$R(B) = \max\{|a(0)|, \psi(a), D_2(a), D_3(a), \ldots, D_m(a), \sup_{s>o} \psi(a; s)\}.$$

13.5 The Action of an Affine Map

The previously described approach is applicable also in the study of weighted shift operators, generated by affine maps, i.e. maps of the form

$$\alpha(x) = Ax + h, \tag{13.5.12}$$

where h is a given vector.

Let S be the previously introduced algebra, corresponding to the compactification by the sphere at infinity.

Lemma 13.5.1 *The algebra S is invariant with respect to any affine map. The prolongation of affine map (13.5.12) on the spaces*

$$X = \tilde{\mathbb{C}}^m = \mathbb{C}^m \coprod S_\infty^{2m-1}$$

acts by the formula

$$\widehat{\alpha}(x) = \begin{cases} Ax + h, & x \in \mathbb{C}^m, \\ \frac{1}{\|Ax\|} Ax, & x \in S^{2m-1}. \end{cases} \tag{13.5.13}$$

Proof An affine map is a composition of a linear map $x \mapsto Ax$ and the translation $x \mapsto x + h$. Due to the invariance of algebra with respect to nonsingular linear map, it is enough to check invariance with respect to the translation $\alpha(x) = x + h$. Let $a \in S$. Then,

$$\lim_{t \to +\infty} a(t\xi + h) = \lim_{t \to +\infty} a(t(\xi + \frac{1}{t}h)) = \lim_{t \to +\infty} a(t\xi) = \tilde{a}(\xi).$$

In particular, this means that the extension of the map $\alpha(x) = x + h$ acts on the sphere at infinity as the identity mapping, so we obtain the formula (13.5.13). □

The properties of a dynamical system, generated by an affine map, depend on the relation between A and the vector h.

Theorem 13.5.2 *For affine transformation*

$$\alpha(x) = Ax + h$$

one has two qualitatively different cases.

1. *The vector $h \in Im(I - A)$. This condition is equivalent to the stipulation that the map has in $\mathbb{C}^m$ at least one fixed point x_0. After the change of variables $y = x - x_0$, this map transforms into a linear, acting by the formula $y \mapsto Ay$. Therefore, the dynamics of such a map, invariant ergodic measures and properties of weighted shift operators are described by the previous Theorems.*
2. *The vector $h \notin Im(I - A)$. In this case, the trajectory of each point from $\mathbb{C}^m$ tends to the sphere at infinity, and all invariant ergodic measures are concentrated on this sphere at infinity.*

Proof 1. Let $h \in Im(I - A)$, i.e. $\forall x_0 \in \mathbb{C}^m$ such that

$$h = x_0 - Ax_0.$$

Then, $Ax_0 + h = Ax_0 + x_0 - Ax_0 = x_0$ and the point x_0 is fixed. In particular, if $\det(I - A) \neq 0$, then the condition $h \in Im(I - A)$ holds for each h, and in this case, there is the unique fixed point $x_0 = (I - A)^{-1}h$.

We denote by T the translation $Tx = x - x_0$, corresponding to the change of coordinates $y = x - x_0$ and we show that $T\alpha T^{-1}y = Ay$. Indeed, by applying successively distinguished maps, we get

$$y \mapsto y + x_0 \mapsto A(y + x_0) + h = Ay + Ax_0 + x_0 - Ax_0 = Ay + x_0 \mapsto Ay.$$

2. Suppose that $h \notin Im(I - A)$. Let $B = A - I$. For a vector h, there is a decomposition $h = u + z$, where $z \in Im B$ is the projection onto $Im B$ and $u \perp Im B$, where $u \neq 0$. Consider a linear functional $f(x) = \langle x, u \rangle$ on $\mathbb{C}^m$. We have $f(x) = 0$ for $x \in Im B$, and

$$f(h) = \langle h, u \rangle = d \neq 0.$$

Moreover, since $|f(x)| \leq \|u\|\|x\|$, the inequality

$$\|x\| \geq \frac{1}{\|u\|}|f(x)|$$

holds. Since $\alpha(x) = x + Bx + h$, the trajectory of the point x has the form $x_n = x_{n-1} + Bx_{n-1} + h$. Therefore,

$$f(x_n) = f(x_{n-1}) + d = f(x_{n-2}) + 2d = \ldots = f(x) + nd \to \infty,$$

and t $\|x_n\| \to \infty$, as required. $\square$

One can further investigate the behavior of the trajectories and indicate explicitly to which point ξ from the sphere at infinity does the trajectory of the point x converge. We shall not do this here, since already obtained information is sufficient for the description of the spectrum of the weighted shift operator generated by an affine map. In case 1 of the Theorem 13.5.2, the description of the spectrum reduces to the previously considered case of a linear map. In case 2, the results are analogous, but in the formula for spectral radius, one has to exclude measures concentrated on $\mathbb{C}^m$, and to keep only measure concentrated on the sphere at infinity. We shall not formulate general statements, but confine ourselves to the consideration of examples, demonstrating the specifics of the case of affine maps.

Example 13.5.3 Suppose that compactification of the space and the matrix A are the same as in the Theorem 13.4.2. We study an affine map $\alpha(x) = Ax + h$, which, in this case, acts by the formula

$$y_1 = \lambda_1 x_1 + h_1, \quad y_2 = \lambda_2 x_2 + h_2,$$

where $0 < \lambda_1 \leq \lambda_2$.

Theorem 13.5.4 *If $a \in S$, $\widehat{a}(x) \neq 0$ for all $x \in X$, then the spectral radius $R(B)$ of the operator*

$$Bu(x) = \sqrt{|\det A|}\, a(x)\, u(Ax + h)$$

in the space $L_2(\mathbb{C}^2)$ are given by the following expressions.
   a. If $\lambda_1 \neq \lambda_2$, $\lambda_k \neq 1$, then

$$R(B) = \max\{|a(\frac{h_1}{1-\lambda_1}, \frac{h_2}{1-\lambda_2})|, \max_{|z_1|=1} |\widehat{a}(z_1, 0)|, \max_{|z_2|=1} |\widehat{a}(0, z_2)|\}.$$

b. If $\lambda_1 = \lambda_2 \neq 1$, then

$$R(B) = \max\{|a(\frac{h_1}{1-\lambda_1}, \frac{h_2}{1-\lambda_2})|, |\widehat{a}(\xi)|, \xi \in S^3_\infty\}.$$

c. If $\lambda_1 < \lambda_2 = 1$ and $h_2 = 0$, then

$$R(B) = \max\{\sup_{x_2 \in \mathbb{C}} |a(\frac{h_1}{1-\lambda_1}, x_2)|, \max_{|z_1|=1} |\widehat{a}(z_1, 0)|, \max_{|z_2|=1} |\widehat{a}(0, z_2)|\}.$$

d. If $\lambda_1 < \lambda_2 = 1$ and $h_2 \neq 0$, then

$$R(B) = \max\{\max_{|z_1|=1} |\widehat{a}(z_1, 0)|, \max_{|z_2|=1} |\widehat{a}(0, z_2)|\}.$$

e. If $\lambda_1 = \lambda_2 = 1$ and $h \neq 0$, then

$$R(B) = \max\{|\widehat{a}(\xi)|, \xi \in S^3_\infty.\}.$$

f. If $\lambda_1 = \lambda_2 = 1$ and $h = 0$, then

$$R(B) = \sup\{|a(x)| : x \in \mathbb{C}^2\} = \max\{|\widehat{a}(\xi)|, \xi \in \mathbb{C}^2 \bigcup S^3_\infty\}.$$

Here the cases **a, b, c** and **f** correspond to the case 1 of the Theorem 13.5.2, the cases **d** and **e** correspond to the case 2.

References

1. Antonevich, A. B. (1988). *Linear functional equations. Operator approach* (English transl. Birkhauser, Basel, Boston, Berlin, 1996). Minsk: Universitetskoe (in Russian).
2. Chicone, C., & Latushkin, Yu. (1999). *Evolution semigroups in dynamical systems and differential equations*. Providence: AMS.
3. Antonevich, A., & Lebedev, A. (1994). *Functional differential equations: I. C*-theory*. Harlow: Longman Scientific & Technical.
4. Antonevich, A., & Lebedev A. (1998). *Functional and functional-differential equations. AC*−algebraic approach* (English transl. in Amer. Math. Soc. Transl. Ser. 2. 1999, pp. 25–116). *Trudy Sankt-Peterburgskogo Matematicheskogo Obshchestva, 6*, 34–140.
5. Gelfond, A. O. (1967). *Finite differences calculus*. Moscow: Nauka.
6. Hale, J. (1977). *Theory of functional differential equations*. New York-Heidelberg-Berlin: Springer.
7. Halmos, P. (1956). *Ergodic theory*. New York: Chelsea.
8. Karapetiants, N., & Samko, S. (2001). *Equations with involutive operators*. Birkhäuser: Boston-Basel-Berlin.

9. Kravchenko, V. G., & Litvinchuk, G. S. (1994). *Introduction to the theory of singular integral operators with shift, mathematics and its applications* (Vol. 289). Dordrecht: Kluwer.

10. Kolmanovsli, V. B., & Nosov, V. R. (1981). *Stability and periodic condition of controlled systems with aftereffects*. Moscow: Nauka.

11. Kornfeld, I. P., Sinai, Y. G., & Fomin, S. V. (1980). *Ergodic theory*. Moscow: Nauka.

12. Katok, A., & Hasselblatt, B. (1998). *Introduction to the modern theory of dynamical systems*. Cambridge: Cambridge University Press.

13. Skubachevskii, A. L. (1997). *Elliptic functional differential equations and applications*. Birkhäuser: Basel-Boston-Berlin.

14. Antonevich, A. B. (1975). On a class of pseudodifferential opearators with deviating argument on the torus. *Differentsial'nye Uravneniya, 11*(9), 1550–1557.

15. Antonevich, A. B. (1979). Operators with a shift generated by the action of a compact Lie group. *Sibirskii Matematicheskii Zhurnal, 20*(3), 467–478.

16. Daniluk, A., & Stochel, J. (1997). Seminormal composition operators induced by affine transformations. *Hokkaido Mathematical Journal, XXVI*(2), 377–404.

17. Stochel, J. (1990). Seminormal composition operators on L^2 spaces induced by matrices. *Hokkaido Mathematical Journal, 19*, 307–324.

Chapter 14
The Subtle Spectral Properties

Along with the description of the spectrum of the operator B, i.e., obtaining conditions of invertibility of operators of the form $B - \lambda I$, some more subtle properties of operators $B - \lambda I$ are interesting in the case when λ belongs to the spectrum. Those are properties such as the closure of the image of the operator, the dimension of kernel and cokernel, the existence of one-sided inverse operator. In applications, these properties enable us to obtain conditions of solvability of corresponding functional of functional-differential equations.

It specifically deals with questions of essential spectra, the subtle spectral properties of an operator generated by a map of the Morse-Smale type, the spectral properties of operators induced by linear maps.

14.1 The Essential Spectra

Recall that the operator B is the Fredholm operator if its kernel ker B and cokernel ker B^* are finite dimensional and the image ImB is closed. Integer number

$$indB = \dim \ker B - \dim \ker B^*$$

is called *index of the Fredholm operator B*.

The operator B is called a Φ^+-operator if its image is closed and the kernel of the adjoint operator is finite dimensional. The operator B is called a Φ^--operator if its image is closed and the kernel is finite dimensional. In particular, all right-sided invertible operators are Φ^--operator.

In another terminology, a description of subtle properties of the operators $B - \lambda I$ is the problem on finding the so-called essential spectra of the operator. Every "good" property of non-invertible operator $B - \lambda I$ splits the spectrum into two parts—the part on where property holds, and the part where property does not hold, being called the corresponding *essential spectrum. The Fredholm spectrum* is used more often

© Springer International Publishing Switzerland 2014
Ć.B. Dolićanin and A.B. Antonevich, *Dynamical Systems Generated by Linear Maps*,
DOI: 10.1007/978-3-319-08228-8_14

$$\Sigma_F(B) = \{\lambda \in \mathbb{C} : B - \lambda I \text{ is not the Fredholm operator}\}.$$

Other forms of essential spectrum of the operator B are also important:

$$\Sigma^+(B) = \{\lambda \in \mathbb{C} : B - \lambda I \text{ does not have a left inverse}\},$$

$$\Sigma^-(B) = \{\lambda \in \mathbb{C} : B - \lambda I \text{ does not have a right inverse}\},$$

$$\Sigma_F^+(B) = \{\lambda \in \mathbb{C} : B - \lambda I \text{ is not a}\Phi^+\text{-operator}\},$$

$$\Sigma_F^-(B) = \{\lambda \in \mathbb{C} : B - \lambda I \text{ is not a}\Phi^-\text{-operator}\},$$

$$\Sigma_K(B) = \{\lambda \in \mathbb{C} : Im(B - \lambda I) \text{ is not closed}\}.$$

It follows from the Theorem 13.1.1, that the form of the spectrum of the weighted shift operator B is uniquely determined by the coefficient a and the set $M_\alpha(X)$; this set is one of the characteristics of the dynamic of the map α.

The task is to find other dynamic properties of the map α, affecting the corresponding subtle properties of the operator $B - \lambda I$.

This problem is solved only for some special types of maps, in particular, for the maps of the Morse-Smale type. In the general case the connection between dynamic properties of the map α and subtle spectral properties of the operator $B - \lambda I$ have not been studied.

14.2   The Subtle Spectral Properties of an Operator Generated by a Map of the Morse-Smale Type

Let $\alpha : X \to X$ be a map of the Morse-Smale type, where all periodic points are fixed points. By the Theorem 13.1.1, for $a(x) \neq 0$, the spectrum of the weighted shift operator B, generated by that map, is the ring

$$\Sigma(B) = \{\lambda \in \mathbb{C} : r(B) \leq |\lambda| \leq R(B)\},$$

where

$$R(B) = \max_{x \in Fix(\alpha)} |a(x)|, \quad r(B) = \min_{x \in Fix(\alpha)} |a(x)|,$$

$Fix(\alpha) = \{F_1, F_2, \ldots, F_q\}$ is the set of fixed points.

Circles

$$S_k = \{\lambda : |\lambda| = |a(F_k)|\}$$

belong to the spectrum $\Sigma(B)$, and they split the spectrum into smaller subrings. The study shows that the properties of the operator $B - \lambda I$ are the same for λ from the same subring, and may be different for λ from different subrings. Therefore, the problem consists of the description of properties of the operator $B - \lambda I$ for different subrings.

In papers [1], the model example of an operators has been studied—the example of an operators induced by a map of Morse-Smale type, having three fixed points F_1, F_2 and F_3, where the point F_1 is repelling, the point F_2 is a saddle point and the point F_3 is attracting. In this example, the spectrum splits into two subrings and the properties of the operator $B - \lambda I$ depend on the validity of an inequality between numbers $|a(F_k)|$ and $|\lambda|$, where the form of these inequalities is determined by the dynamics of the map α. Even in that relatively simple case, the formulation of results is very cumbersome—it involves 18 different subcases.

It turns out that the subtle properties of the operator $B - \lambda I$ have different dependence on the value $a(F_k)$ in fixed points of various types. This is a significant difference from the description of the spectrum, since in the previous description of the spectrum, the fixed points are of equal importance—the dependence on the value of coefficients at different fixed points in the same.

The dynamic characteristics of the mapping α, affecting the considered property of the operators, were discovered in [2, 3]; this allowed one to obtain a simple compact foreseeable solution to the problem.

That dynamic characteristic of the map α of the Morse-Smale type which was useful in obtaining solutions of the problem in question turned out to be an *oriented graph* G_α, describing the dynamics of the map. The vertices of the graph G_α are fixed points F_k. An *oriented edge* $F_k \to F_j$ is included in the graph, if and only if, there is a point $x \in X$ such that its trajectory tends to F_j as $n \to +\infty$ and tends to F_k as $n \to -\infty$.

The discussed properties may be restated in another form. We denote by X_k^+ (X_k^-) *positive (negative) pool* of the point F_k—the set of those points x whose trajectories tend to F_k as $n \to +\infty$ ($n \to -\infty$).

In these terms, the oriented edge $F_k \to F_j$ is included in the graph if and only if

$$X_j^+ \bigcap X_k^- \neq \emptyset.$$

A map α of the Morse-Smale type is called *irreducible* if there exist k and j such that the set

$$X_j^+ \bigcap X_k^-$$

is dense in X.

The use of the introduced graph permitted one to obtain a simple formulation of the important properties of the weighted shift operators.

By using the coefficient a and number λ, one forms two subsets of the set of vertices of the graph

$$G^+(a, \lambda) = \{F_k \in Fix(\alpha) : |a(F_k)| > |\lambda|\}, \qquad (14.2.1)$$

$$G^-(a, \lambda) = \{F_k \in Fix(\alpha) : |a(F_k)| < |\lambda|\}.$$

It is clear that

$$G^+(a, \lambda) \bigcap G^-(a, \lambda) = \emptyset.$$

We say that subsets $G^+(a, \lambda)$ and $G^-(a, \lambda)$ give a *decomposition of the graph*, if the condition

$$G_\alpha = G^+(a, \lambda) \bigcup G^-(a, \lambda)$$

holds, i.e. if

$$|a(F_k)| \neq |\lambda| \quad \text{for all} \quad F_k \in Fix(\alpha).$$

The graph decomposition shall be called *oriented to the right*, if any edge connecting the point $F_k \in G^-(a, \lambda)$ to the point $F_j \in G^+(a, \lambda)$, is oriented from F_k to F_j.

The decomposition shall be called *oriented to the left*, if any edge connecting the point $F_k \in G^-(a, \lambda)$ to the point $F_j \in G^+(a, \lambda)$, is oriented from F_j to F_k.

The basic result in this direction is the following (see [3, 4]).

Theorem 14.2.1 *Let α be a map of the Morse-Smale type of the compact space X, $a \in C(X)$ and $B = aT_\alpha$ be weighted shift operator. The operator $B - \lambda I$ is invertible from the right (left), if and only if, the subsets $G^+(a, \lambda)$ and $G^-(a, \lambda)$ form a decomposition of the graph G_α being oriented to the right (to the left).*

If α is irreducible, then the image of the operator is closed if and only if the operator is one-sided invertible.

From this Theorem, it is easy to describe all previously mentioned essential spectra of the operator.

14.3   The Subtle Spectral Properties of Operators Induced by Linear Maps

We study operators of the form (13.2.4), in the space $L_2(\mathbb{C}^m)$ and $L_2(\mathbb{R}^m)$ with coefficients, corresponding to the two simplest compactification.

14.3.1 One-Point Compactification

According to the Theorem 5.1.1, the prolongation (13.3.7) on the one-point compactification is a map of the Morse-Smale type if and only if the moduli of all eigenvalues are different from 1. Then, the set of fixed points consists of two points 0 and ∞, and the spectrum is a ring with radiuses $|a(0)|$ and $|a(\infty)|$.

To apply the Theorem 14.2.1, we have to construct an oriented graph with two vertices 0 and ∞, corresponding to the map in question. For this, we have to find the sets $X^{\pm}(0)$ and $X^{\pm}(\infty)$.

In fact, such a description has been obtained in the Theorem 5.1.1. Let λ_k be eigenvalues of the operator A and suppose that $|\lambda_k| \neq 1$ for all k. There are three possible subcases:

(i) $|\lambda_k| < 1$ for all k.
(ii) $|\lambda_k| > 1$ for all k.
(iii) $|\lambda_k| < 1$ for $k \leq k_0$; $|\lambda_k| > 1$ for $k > k_0$.

We consider the behavior of the trajectory, and we construct the graph G_α, in these three cases.

(i) trajectories of all points from $\mathbb{R}^m$ tends to the point 0, when $n \to +\infty$ and tend to ∞ for $n \to -\infty$, i.e.

$$X^+(0) = \mathbb{C}^m, \; X^-(0) = \{0\},$$

$$X^+(\infty) = \{\infty\}, X^-(\infty) = \mathbb{C}^m,$$

and the graph contains only one oriented edge: $\infty \to 0$.
(ii) we obtain a graph with the opposite orientation—the graph contains of only one oriented edge: $0 \to \infty$.
(iii) all four sets $X^{\pm}(0)$ are $X^{\pm}(\infty)$ nontrivial. Namely,

$$X^+(0) = \{x \in \mathbb{C}^m : r(k(x)) < 1\} = M(k_0);$$

$$X^+(\infty) = \mathbb{C}^m \setminus M(k_0).$$

$$X^-(0) = \{x \in \mathbb{R}^m : |\lambda(k(x))| < 1\} = L'(k_0).$$

$$X^-(\infty) = \mathbb{R}^m \setminus L'(k_0).$$

Therefore, the graph G_α contains two oppositely oriented edges: $\infty \to 0$ and $0 \to \infty$.

Now, by applying the Theorem 14.2.1, we obtain a description of the properties of the operator $B - \lambda I$.

Theorem 14.3.1 *Let the moduli of all eigenvalues of the operator A are different from 1 and the coefficient a is a continuous function having the limit $a(\infty)$ at infinity.*

(i) *Suppose $|\lambda_k| < 1$ for all k.*

If $|a(\infty)| < |\lambda| < |a(0)|$, then the operator $B - \lambda I$ is invertible from the right.

If $|a(0)| < |\lambda| < |a(\infty)|$, then the operator $B - \lambda I$ is invertible from the left.

(ii) *Suppose $|\lambda_k| > 1$ for all k.*

If $|a(\infty)| < |\lambda| < |a(0)|$, then the operator $B - \lambda I$ is invertible from the left.

If $|a(0)| < |\lambda| < |a(\infty)|$, then the operator $B - \lambda I$ is invertible from the right.

(iii) *If $|\lambda_k| < 1$ for $k \leq k_0$ and $|\lambda_k| > 1$ for $k > k_0$, then for all $\lambda \in \Sigma(B)$ the image of the operator $B - \lambda I$ is not closed.*

(iv) *If $|\lambda| = |a(0)|$ and if $|\lambda| = |a(\infty)|$, then the image of the operator $B - \lambda I$ is not closed.*

14.3.2 The Compactification by the Sphere at Infinity

We discuss the compactification by the sphere at infinity. In the complex case, an extension α of a linear map A to such a compactification cannot be a map of the Morse-Smale type, since if there is a fixed point, then there is a whole circle of fixed points. But, in the case $\mathbb{R}^m$, the prolongation of the form (13.4.10) may be a map of the Morse-Smale type. In particular, α is a map of the Morse-Smale type such that all periodic points are fixed, if and only if, all eigenvalues are positive, different, different from 1 and every eigenvalues has multiplicity 1. As an example, we discuss a diagonal matrix

$$A = diag(\lambda_1, \lambda_2, \dots, \lambda_m),$$

where

$$0 < \lambda_1 < \lambda_2 < \cdots < \lambda_{k_0} < 1 < \lambda_{k_0+1} < \cdots < \lambda_m.$$

Corresponding map α is a map of the Morse-Smale type having $2^m + 1$ fixed points: the point $0 \in \mathbb{R}^m$ and points $\pm e(k) \in S_\infty^{m-1}$.

In order to apply the Theorem 14.2.1, we have to construct an oriented graph with vertices 0 and $\pm e(k)$, corresponding to the given map. The necessary information is contained in Theorems describing the dynamics. For different k_0, we obtain different graphs and, consequently, different condition on one-sided invertibility.

In the given example, it is possible to simplify the problem. The space $\mathbb{R}^m$ is decomposed into 2^m polyhedral angles of the form

$$\{x \in \mathbb{R}^m : \pm x_k \geq 0\}$$

invariant with respect to the action A. Therefore, the operator B decomposes into a direct sum of operators given by the same formula on the corresponding polyhedral angle, and so it is enough to study the operator in each of these polyhedral angles. The properties of these operators are analogous, so we discuss one of them. Namely, let us consider the operator B^+, acting on the space $L_2(\mathbb{R}^m_+)$. Let X^+ be the closure of the set $\mathbb{R}^m_+$ on X. The fixed points from X^+ are 0, $e(1), e(2), \ldots, e(m)$, and one has to construct the graph $G(\alpha)$ with these vertices, describing the dynamics of the map (13.4.10) on X^+. This is easy to do using the Theorems 5.1.1 and 5.3.2.

Theorem 14.3.2 *Suppose the previously given hypotheses are satisfied and let k_0 be the largest of indices of eigenvalues, for which $\lambda_k < 1$. Then the map (13.4.10) is irreducible on X^+ and the corresponding graph $G(\alpha)$ has the form*

$$e(1) \to e(2) \to e(k_0) \to 0 \to e(k_0 + 1) \to \cdots \to e(k_m).$$

All possible coefficients a and number λ may induce $2^{m+1} - 2$ different nontrivial decompositions of the form (14.2.1), but oriented at the right among them are only m decompositions. To be exact, a nontrivial decomposition is oriented at the right iff $G^-(a, \lambda)$ is one of the m following sets:

$$\{e(1), e(2), \ldots, e(j)\}, 1 \le j \le k_0;$$

$$\{e(1), e(2), \ldots, e(k_0), 0\},$$

$$\{e(1), e(2), \ldots, e(k_0), 0, e(k_0 + 1), e(j)\}, k_0 < j < m.$$

Theorems 14.3.2 and 14.2.1 allow to describe the properties of the operator $B - \lambda I$ for all spectral values λ and, in particular, find explicitly all essential spectra.

Example 14.3.3 Let us look the diagonal matrix $A = diag(1/3, 1/2, 3)$ and consider the operator B in $L_2(\mathbb{R}^3_+)$, given by expression

$$Bu(x) = \frac{2 + x_1 + 3x_2 + 4x_3}{1 + \sqrt{x_1^2 + x_2^2 + x_3^2}} u(Ax). \qquad (14.3.1)$$

The prolongation α of the form (13.4.9) has four fixed points:

$$0 \in \mathbb{R}^3_+; \quad e(1), e(2), e(3) \in S^2_\infty.$$

The oriented graph $G(\alpha)$, describing the dynamics of α on X^+, is

$$e(1) \to e(2) \to 0 \to e(3).$$

The reduced coefficient is

$$a(x) = \sqrt{2}\,\frac{2 + x_1 + 3x_2 + 4x_3}{1 + \sqrt{x_1^2 + x_2^2 + x_3^2}},$$

and the condition (13.4.9) holds. The values of the reduced coefficient at the fixed points are

$$a(0) = 2\sqrt{2}, \ a(e(1)) = \sqrt{2}, \ a(e(2)) = 3\sqrt{2}, \ a(e(3)) = 4\sqrt{2}.$$

The previous results give us the description of the properties of the operator $B - \lambda I$ for all spectral values λ.

Theorem 14.3.3 *The spectrum of the operator* (14.3.1) *is*

$$\Sigma(B) = \{\lambda : \sqrt{2} \leq |\lambda| \leq 4\sqrt{2}\}.$$

If $\sqrt{2} < |\lambda| < 2\sqrt{2}$ *and if* $3\sqrt{2} < |\lambda| < 4\sqrt{2}$, *then the operator* $B - \lambda I$ *is invertible from the right.*
For all the rest spectral values λ, *the image of* $B - \lambda I$ *is not closed.*

References

1. Antonevich, A., & Jakubowska, J. (2010). Weighted translation operators generated by mappings with saddle points: A model class. *Journal of Mathematical Sciences, 164*(4), 497–517.
2. Antonevich, A., & Makowska, Yu. (2008). On spectral properties of weighted shift operators generated by mappings with saddle points. *Complex analysis and Operator theory, 2*, 215–240.
3. Antonevich, A., & Makowska, Yu. (2009). One-sided invertibility of weighted shift operator. In V. A. Sadovnichy (Ed.) *Materials of the International conference "Modern Problems of Mathematics, Mechanics and their Applications", Dedicated to the 70-th Anniversary of Rector of MSU Academy* (p. 16), Moscow, March 30–April 02.
4. Makowska, J. (2010). One-side invertibility of the weighted shift operators. In P. Kielanowski, V. Buchshtaber, A. Odzijewicz, M. Shlichenmaier, & T. Voronov (Eds.), *XXIX Workshop on Geometric Methods in Physics* (P. 101–105). American Institute of Physics, Melville, New York.

Chapter 15
The Weighted Shift Operators in Spaces of Vector Functions

In spaces of vector functions, on an arbitrary set X, a weighted shift operator has the same form as in the scalar case:

$$Bu(x) = a_0(x)u(\alpha(x)), \quad x \in X, \tag{15.0.1}$$

where $a_0(x)$ is not a scalar but a matrix-valued function on X.

Such operators have much more complex structure, than in spaces of scalar functions and the results have less explicit character. In particular, for such operators, the results are usually formulated with the help of auxiliary objects, which, in the general case, are not constructed explicitly. It is precisely in the construction and the investigation of these auxiliary objects that an information concerning the behavior of subspace trajectories is needed.

We, here, discuss only the problem concerning investigation of spectral properties of operators. Note that an information on the behavior of subspace trajectories is needed in other problems. Among them, for example, in *the problem on the normal form* of weighted shift operators.

Let $B = aT_\alpha$ be a weighted shift operator, and $S(x)$ be a continuous invertible matrix-function; then $S(x)BS(x)^{-1}$ is a weighted shift operator of the form $\tilde{a}T_\alpha$ with the new coefficient

$$\tilde{a}(x) = S(x)a(x)S^{-1}(\alpha(x)).$$

The problem consists in the clarification what is the simplest form to which the coefficient of a weighted shift operator may be reduced by these transformations. This problem is analogous to the problem on the reduction of a matrix to a canonical form; if one solves it, one obtains an analogue of the Jordan form for weighted shift operators.

In particular, if the coefficient of an operator B may be reduced to a diagonal form, then a study of this operator with a matrix coefficient can be reduced to the study of m operators with scalar coefficients, which substantially simplifies a description of the properties of B.

© Springer International Publishing Switzerland 2014
Ć.B. Dolićanin and A.B. Antonevich, *Dynamical Systems Generated by Linear Maps*,
DOI: 10.1007/978-3-319-08228-8_15

15.1 General Approaches

For the study of the spectrum of a weighted shift operator in a space of vector-functions, one may apply two approaches:

hyperbolic, using the relation of the operator to the hyperbolic property of the associated linear extension of a dynamical system, and

trajectorial, based on the relation to a family of discrete operators, induced by restrictions on trajectories of a dynamical system. Considering the history of the problem and the development of these approaches, see monographs [1–3] and surveys [4, 5].

Hyperbolic approach. Associated linear extension β of the map α is constructed as follows.

The product $E = X \times \mathbb{C}^m$ may be seen as a vector bundle over X with the natural projection; the fibre E_x over the point $x \in X$ is the set

$$E_x = \{(x, \xi) : \xi \in \mathbb{C}^m\} = \{x\} \times \mathbb{C}^m.$$

We define a map $\beta : E \to E$ by the formula

$$\beta(x, \xi) = (\alpha(x), [a(x)]^{-1}\xi), \quad x \in X, \ \xi \in \mathbb{C}^m.$$

The map β is continuous, and it maps linearly the fibre E_x onto the fiber $E_{\alpha(x)}$. This means that β is a *linear extension of the map α* (see, e.g. [6]).

For powers, i.e. iterations, of the map β, we have

$$\beta^n(x, \xi) = (\alpha^n(x), a(x; n)\xi), n \in \mathbb{Z},$$

where

$$a(x; n) = [a(\alpha_{n-1}(x))]^{-1} \times [a(\alpha_{n-2}(x))]^{-1} \times \cdots \times [a(\alpha(x))]^{-1} \times [a(x)]^{-1}$$

is the so-called *cocycle*, corresponding to β.

We introduce notation: $\|(x, \xi)\| = \|\xi\|$, where $\|\xi\|$ is a norm on $\mathbb{C}^m$.

The subbundle E^s of the bundle E is called *stable in the positive direction* for the linear extension β, if it is invariant with respect to β, and there are constants γ, $0 < \gamma < 1$, and $C > 0$, for which the following inequality holds:

$$\|\beta^n(x, \xi)\| \le C\gamma^n \|\xi\|, \quad x \in X, \ (x, \xi) \in E^s, \ n = 1, 2, \ldots \qquad (15.1.2)$$

The subbundle E^u is called *unstable in the positive direction*, if there are constants γ, $0 < \gamma < 1$, and $C > 0$, so that

$$\|\beta^n(x, \xi)\| \ge C\gamma^{-n} \|\xi\|, \quad x \in X, \ (x, \xi) \in E^u, \ n = 1, 2, \ldots \qquad (15.1.3)$$

A linear extension is *hyperbolic*, if the vector bundle E is split into a direct Whitney sum of stable and unstable subbundle [6]:

$$E = E^s \bigoplus E^u.$$

A *dynamical spectrum* $\Sigma(\beta)$ of the linear extension β is the set of numbers λ, where the linear extension β_λ from the spectral family of linear extensions

$$\beta_\lambda(x, \xi) = (\alpha(x), \lambda[a(x)]^{-1}\xi), \ \lambda \in \mathbb{C}, \lambda \neq 0, \tag{15.1.4}$$

is not hyperbolic.

A linear operator A is called *hyperbolic*, if its spectrum does not intersect the unit circle.

Proposition 15.1.1 (Hyperbolicity Theorem) *Let X be a compact space, μ—a measure on X, whose support coincides with the whole space, $\alpha : X \to X$—an invertible continuous map, preserving the class of the measure μ, the set of non-periodic points of the map α is dense in X and the space X is α-irreducible. Let also a be a continuous matrix-valued function and $\det a(x) \neq 0$ for all x.*

The operator $B - l$ in the space of vector functions $L_2(X, \mathbb{C}^m, \mu)$ is invertible if and only if the associated linear extension β is hyperbolic, and under this condition the operator B is hyperbolic.

Corollary 15.1.2 *The spectrum of the operator B coincides with the dynamic spectrum of the linear extension β, and it consists of most m disjoint rings with the center at the point 0.*

A *trajectorial approach* is based on the following construction.

Let $l_2(\mathbb{Z}, \mathbb{C}^m)$ be the space of two-sided sequences $u = (u(k)), k \in \mathbb{Z}, u(k) \in \mathbb{C}^m$, with the norm $\|u\| = (\sum_k \|u(k)\|^2)^{1/2}$. Denote by T a shift operator, acting on $l_2(\mathbb{Z}, \mathbb{C}^m)$, by the formula

$$(Tu)(k) = u(k + 1).$$

For each $\tau \in X$ we construct, using the coefficient $a \in C(X)$, a sequence of matrix $a_\tau(k) = a(\alpha^k(\tau))$. Denote by $B(\tau), \tau \in X$, a discrete weighted shift operator $a_\tau T$, acting on $l_2(\mathbb{Z}, \mathbb{C}^m)$, i.e. the operator given by the formula

$$(B(\tau)u)(k) = a(\alpha^k(\tau))u(k + 1). \tag{15.1.5}$$

The operator $B(\tau)$ is constructed by the restriction of the coefficient a to the trajectory of the point τ, which clarifies the term "trajectorial approach".

Proposition 15.1.3 (Discretization Theorem) *Under the conditions of the Proposition 15.1.1, the operator $B - l$ is invertible if and only if all operators $B(\tau) - I, \ \tau \in X$, are invertible.*

Corollary 15.1.4

$$\sigma(B) = \bigcup_{\tau \in X} \sigma(B(\tau)).$$

The formulated proposition reduces the problem of the description of the spectrum to other problems, but for an arbitrary operator, solutions of these auxiliary problems cannot be obtained in an explicit form. Therefore, it is interesting to find special cases where one may obtain more explicit results, for example, the explicit conditions of hyperbolicity or explicit description of spectrum for discrete weighted shift operators.

Let us first discuss an elementary example.

Proposition 15.1.5 *Let B be a weighted shift operator whose reduced matrix coefficient is constant: $a(x) \equiv A$. Then,*

$$\sigma(B) = \bigcup_k S_k,$$

where S_k is the circle of radius $r(k)$:

$$S_k = \{\lambda : |\lambda| = r(k)\},$$

and numbers r_k are moduli of eigenvalues of the matrix A.

Proof We apply the Corollary 15.1.2. In the given case, we have a simple form of the cocycle, corresponding to β_λ from the spectral family of linear extensions (15.1.4):

$$a(x, n) = \lambda^n A^{-n}.$$

Therefore, the investigation of the linear extension β_λ is reduced to the investigation of powers of the matrix A. We can apply the previously given investigation of the behavior of trajectories under the action of powers of A. Inequalities (15.1.2) hold for vector $\xi \in \mathbb{C}^m$ if and only if

$$\xi \in \bigoplus_{r(k)>|\lambda|} L(k) = L_\lambda^s.$$

Therefore, the maximal stable subbundle is the product bundle

$$E^s = \{(x, \xi) : x \in X, \xi \in L_\lambda^s\}.$$

Analogously, inequalities (15.1.3) hold if and only if

$$\xi \in \bigoplus_{r(k)<|\lambda|} L(k) = L_\lambda^u$$

and the maximal unstable subbundle is the product bundle

$$E^u = \{(x, \xi) : x \in X, \xi \in L_\lambda^u\}.$$

Therefore, the equality $E = E^s \oplus E^u$ holds if and only if $r(k) \neq 1$ for all k, and this condition is equivalent to the invertibility of the operator $B - \lambda I$.   □

An analogous result is obtained in the case when the coefficient is constant on each trajectory: the condition $a(\alpha(x)) = a(x)$ holds. Let $r(k, x)$ be moduli of eigenvalues of the matrix $a(x)$. In that case,

$$\sigma(B) = \bigcup_x \bigcup_k \{\lambda : |\lambda| = r(k, x)\}.$$

If the coefficient does not commute with the shift, the situation is more complex. For example, the following proposition holds.

Proposition 15.1.6 *Let α be a map of the Morse-Smale type, all periodic points $F(j)$ are fixed points of α. Denote by $r(k, j)$ the moduli of eigenvalues of the matrix $a(F(j))$, and let*

$$R = \max_{k,j} r(k, j), \quad r = \min_{k,j} r(k, j).$$

Then,

$$\sigma(B) \subset K = \{\lambda : r \leq |\lambda| \leq R\},$$

where all circles

$$S_{k,j} = \{\lambda : |\lambda| = r(k, j)\}$$

belong to the spectrum.

The set $K \setminus \bigcup_{k,j} S_{k,j}$ is the union of finitely many disjoint open rings, some of them belonging to the spectrum, and others do not. To determine which of these open rings belong to the spectrum, requires a further analysis.

By the trajectorial approach, the problem can be reduced to the consideration of discrete weighted shift operators (15.1.5), and in the case of a map of the Morse-Smale type, the coefficient a_τ determining a discrete operator (15.1.5), has limits at $\pm\infty$. Let us show in an example, how one uses information concerning behavior of a trajectory under a linear map, in the analysis of such discrete operators.

15.2 A Spectrum and Subtle Spectral Properties
of the Model Operator with Matrix Coefficients

Consider a discrete weighted shift operator

$$Bu(k) = a(k)u(k+1),$$

acting on the space $l_2(\mathbb{Z}, \mathbb{C}^m)$, whose coefficient satisfy the condition: there are limits

$$\lim_{k \to \pm\infty} a(k) := A_\pm.$$

If as X we take compactification of the space $\mathbb{Z}$ by points $\pm\infty$, then the prolongation α of the map $k \to k+1$ on X is a map of the Morse-Smale type, having two fixed points $\pm\infty$. Therefore, one can apply the Proposition 15.1.6, where this proposition may be strengthened for the case of discrete operators.

Proposition 15.2.1 *Let $r(k)_\pm$ be moduli of eigenvalues of the $A_\pm$, and*

$$R = \max r(k)_\pm, \quad r = \min r(k)_\pm.$$

Then,

$$\sigma(B) \subset K = \{\lambda : r \le |\lambda| \le R\}.$$

Operator $B - \lambda I$ is Fredholm if and only if

$$\lambda \notin \Sigma_F(B) = \bigcup S_k^\pm,$$

where

$$S_k^\pm = \{\lambda : |\lambda| = r(k)_\pm\}.$$

For $\lambda \in \Sigma_F(B)$, the subspaces $\ker(B-\lambda I)$ and $\ker([B-\lambda I]^*)$ are finite dimensional, and the image $Im(B - \lambda I)$ is not closed.

In the general case, even for this class of discrete weighted shift operators the answer to the question which of the rings forming the set $K \setminus \Sigma_F(B)$ belong to the spectrum, cannot be obtained in an explicit form. But, in some special cases, it is possible to give an explicit description of the spectrum of the operator B, using previously described results on the behavior of vectors and subspaces under an action of a linear operator.

Let us restrict ourselves here to the analysis of one model example. Consider discrete weighted shift operator B, where the coefficient has only two values, and has the form

$$a(k) = \begin{cases} A_+, & k \geq 0, \\ A_-, & k < 0. \end{cases} \tag{15.2.6}$$

Since $B - \lambda I = \lambda(\frac{1}{\lambda}B - I)$, it is enough to consider operator of the form $B - I$ with an arbitrary coefficient from the mentioned class.

If for some k, one has $r(k)_+ = 1$ or $r(k)_- = 1$, then the properties of the operator $B - I$ are described in the Proposition 15.2.1: subspaces $\ker(B - \lambda I)$ and $\ker([B - \lambda I]^*)$ are finite dimensional, and the image $Im(B - \lambda I)$ is not closed. Therefore, one only has to discuss the case when $r(k)_\pm \neq 1$ for all k.

In the given example, the linear extension $\beta : X \times \mathbb{C}^m \to X \times \mathbb{C}^m$ acts by the formula

$$\beta(k, \xi) = \begin{cases} (k+1, [A_+]^{-1}\xi), & k \geq 0, \\ (k+1, [A_-]^{-1}\xi), & k < 0. \end{cases} \tag{15.2.7}$$

Each subbundle invariant with respect to β, is uniquely determined by the fiber over the point 0, i.e. by a subspace $V \subset \mathbb{C}^m$. Let us clarify what conditions must the subspace V satisfy in order to induce a stable subbundle E^s, so that $V = E_0^s$.

By $L(k)_\pm$ we denote subspaces in $\mathbb{C}^m$, corresponding to $r(k)_\pm$ and giving decompositions of the form (12.1.1) for $A_\pm$. Let

$$L_+^s = \bigoplus_{r(k)_+ > 1} L(k)_+, \quad L_+^u = \bigoplus_{r(k)_+ < 1} L(k)_+;$$

$$L_-^s = \bigoplus_{r(k)_- > 1} L(k)_-, \quad L_-^u = \bigoplus_{r(k)_- < 1} L(k)_-.$$

Lemma 15.2.2 *A subspace $V \subset \mathbb{C}^m$ induces a stable subbundle if and only if the following conditions hold:*

(i) $V \subset L_+^s$;
(ii) $V \cap L_-^u = \{0\}$;
(iii) *there exist limits $\lim A_+^{-n}(V)$ and $\lim A_-^n(V)$ as $n \to +\infty$.*

Proof Since the action of β on the vector $\xi \in E_0^s$ reduces to the multiplication by powers of the matrix $[A_+]^{-1}$, a necessary condition for inequality (15.1.2) is, according to the Theorem 5.1.1, inclusion

$$V \subset L_+^s.$$

Let us now consider the point $(-n, \eta)$, from the fiber E_{-n}^s, where $n > 0, \eta \neq 0$. We have

$$\beta^n(-n, \eta) = (0, A_-^{-n}\eta) = (0, \xi), \; \xi \in V = E_0^s,$$

192 15 The Weighted Shift Operators in Spaces of Vector Functions

so $\eta = A_-^n \xi$. From (15.1.2) we obtain inequality

$$\|\xi\| = \|A_-^{-n}\eta\| \leq C\gamma^n \|\eta\|,$$

being equivalent to the inequality

$$\|A_-^n \xi\| \geq \frac{1}{C} \frac{1}{\gamma^n} \|\xi\|.$$

According to the description of vector trajectories (Theorem 5.1.1), such an inequality holds if and only if $\xi \notin L_-^u$, and we obtain the second condition: $V \cap L_-^u = \{0\}$.

In addition to that, the fibers of a subbundle, by definition, depend continuously on the point x on all X, including points $\pm\infty$. Therefore, a necessary condition is also the existence of the limit of trajectories of subspaces:

$$E_{+\infty}^s = \lim_{n\to+\infty} (A^+)^{-n}(V), \quad E_{-\infty}^s = \lim_{n\to+\infty} (A_-^n(V).$$

The condition is sufficient following easily from the previous. □

By applying this Lemma to the inverse map β^{-1}, we obtain

Corollary 15.2.3 *A subspace W induces unstable subbundle if and only if the following conditions hold:*
(i) $W \subset L_-^u$;
(ii) $W \cap L_+^s = \{0\}$;
(iii) there are limits: $\lim A_+^{-n}(W)$ and $\lim A_-^n(W)$ as $n \to +\infty$.

In this way, in the construction of invariant subbundles, previously mentioned problems appear concerning the description of subspace trajectories, and determination of the limits of trajectories.

We can now obtain condition for the hyperbolicity of β, being the conditions on the invertibility of the operator $B - I$.

Theorem 15.2.4 *Linear extension β, given by the expression (15.2.7), is hyperbolic if and only if the following conditions hold:*

(i)

$$r(k)_\pm \neq 1;$$

(ii)

$$L_+^s \cap L_-^u = \{0\}, \quad L_-^s \cap L_+^u = \{0\}. \tag{15.2.8}$$

Then, the stable subbundle is induced by the subspace $V = L_+^s$, and unstable is induced by the subspace $W = L_-^u$.

Proof Necessity Let β be hyperbolic and $V = E_0^s$, $W = E_0^u$. It follows from the hyperbolicity, that

$$\dim V + \dim W = m.$$

It follows from the Lemma 15.2.2 and its Corollary, that

$$\dim V \leq \dim L_+^s, \quad \dim W \leq \dim L_-^u.$$

These conditions hold, only if

$$\dim V = \dim L_+^s, \quad \dim W = \dim L_-^u.$$

Since $V \subset L_+^s$, the equality $\dim V = \dim L_+^s$ is possible only when $V = L_+^s$. Similarly, the equality $\dim W = \dim L_-^u$ is possible only when $W = L_-^u$.

Sufficiency Suppose that the conditions (15.2.8) hold. We set $V = L_+^s$. The subspace L_+^s is invariant under the action of A_+ and it is obvious that there is the limit

$$\lim_{n \to +\infty} (A_+)^{-n}(V) = L_+^s.$$

Under the action of the operator A_-, the subspace V is not invariant, and one has to check that there are $\lim A_-^n(V)$. One can apply the Theorem 12.3.2 here, where the existence of the required limit and equality follows

$$\lim_{n \to +\infty} A_-^n(V) = L_-^s.$$

In this way, the subspace $V = L_+^s$ induces stable subbundle E^s.

We analogously obtain that the subspace $W = L_-^u$, if the conditions of the Theorem hold, induces unstable subbundle E^u.

By the construction at each point $k \in X$, (including $\pm\infty$), the following conditions hold:

$$E_k^s \bigcap E_k^u = 0, \quad \dim E_k^s + \dim E_k^u = m,$$

and they equivalent to the required equality

$$E = E^s \bigoplus E^u. \qquad \square$$

If we apply this Theorem to the operator $\frac{1}{\lambda}B - I$, we can find those $\lambda \in K \setminus \Sigma_F(B)$, for which the operator $B - \lambda I$ is invertible.

The approach used above allows not only to obtain conditions on the invertibility of the operator $B - I$, but to investigate this operator in more details in the case of its non-invertibility.

Let

$$d_\pm^s = \dim L_\pm^s, \quad d_\pm^u = \dim L_\pm^u, \quad d = \dim L_+^s \bigcap L_-^u.$$

Note that if $r(k)_\pm \neq 1$, then $d_\pm^u = m - d_\pm^s$.

Theorem 15.2.5 *Under the conditions* $r(k)_\pm \neq 1$, *the operator* $B - I$ *is Fredholm, and*

$$\dim \ker(B - I) = d = \dim[L_+^s \bigcap L_-^u],$$

$$\dim \ker(B^* - I) = d_-^s - d_+^s + d.$$

$$ind(B - I) == d_+^s - d_-^s.$$

Proof By the construction of the map β, the solution of the equation $\xi_k = a(k)\xi_{k+1}$ in the space of all sequences of vectors ξ_k is the trajectory of the vector $(0, \xi_0)$ under the action of β. In order that this sequence belongs to $l_2(\mathbb{Z}, \mathbb{C}^m)$ it is necessary that $\xi_k \to 0$ as $k \to \pm\infty$.

In order for this condition to be fulfilled for $k \to +\infty$, it is necessary that $\xi_0 \in L_+^s$.
In order for this condition to be fulfilled for $k \to -\infty$, it is necessary that $\xi_0 \in L_-^s$.
Therefore, the sequende ξ_k may belong to $\ker(B - I)$ only under the condition

$$\xi_0 \in V = L_+^s \bigcap L_-^u.$$

In this case, for $\xi_o \in V$, the sequence ξ_k decreases exponentially and belongs to the space $l_2(\mathbb{Z}, \mathbb{C}^m)$.

In this way, the sequence ξ_k belongs to $\ker(B - I)$ if and only if $\xi_0 \in V$. It follows from this, that $\dim \ker(B - I) = \dim V$.

The adjoint operator B^* acts on the sequence $\xi = (\xi_k)$ by the formula

$$[B^*\xi]_k = [a(k - 1)]^*\xi_{k-1},$$

it is the operator of the same form as B, but with the shift acting in the opposite direction. Therefore, for the operator B^* the analogous statements hold, where one only has to switch places of $+\infty$ and $-\infty$.

By $L_\pm^*(k)$, we denote subspaces giving a decomposition corresponding to the operator $A_\pm^*$, and let

$$L_+^{s*} = \bigoplus_{r(k)_+ < 1} L(k)_+^*, \quad L_+^{u*} = \bigoplus_{r(k)_+ > 1} L(k)_+^*;$$

$$L_-^{s*} = \bigoplus_{r(k)_- < 1} L(k)_-^*, \quad L_-^{u*} = \bigoplus_{r(k)_- > 1} L(k)_-^*.$$

Then, from the previous discussion, we obtain that

$$\dim \ker(B^* - I) = \dim(L_-^{s*} \cap L_+^{u*}).$$

Let us find that dimension. Let $P_\pm(k)$ be the projector onto $L(k)_\pm$. By the construction, the projectors $P_-(k)$ commute with the operator A_-, the projectors $P_+(k)$ commute with the operator A_+ and one has

$$\sum_k P_\pm(k) = I.$$

The adjoint operator $P_\pm^*(k)$ is a projector onto a certain subspace. These projectors commute with the operator $A_\pm^*$ respectively, and they induce corresponding decomposition of this operator. Therefore, the operator $P_+^*(k)$ projects onto the subspace $L_+^*(k)$, and $P_-^*(k)$ projects onto the subspace $L_-^*(k)$.

From this, we obtain that the operator

$$P_+^{u*} = \bigoplus_{r(k)_+ > 1} P_+(k)^*$$

is the projector onto the subspace L_+^{u*}, and the operator

$$P_-^{s*} = \bigoplus_{r(k)_+ < 1} P_-(k)^*$$

is the projector onto L_-^{s*}.

As it is known, the image of a linear operator in a finite dimensional space is the orthogonal complement to the kernel of the adjoint operator. From this, we obtain that

$$L_+^{u*} = (L_+^s)^\perp, \; L_-^{s*} = (L_-^u)^\perp. \tag{15.2.9}$$

Therefore, there is a relation between dimensions of subspaces

$$V = L_+^s \cap L_-^u \; \text{ and } \; W = L_-^{s*} \cap L_+^{u*}.$$

By (15.2.9), the condition $\xi \in L_+^{u*}$ may be written in the form of d_+^s linearly independent conditions on the coordinates of the vector ξ, the condition $\xi \in L_-^{s*}$ is written as $d_-^u = m - d_-^s$ linearly independent conditions on the coordinates of the vector ξ. But, d of these conditions, are consequences of the conditions from the first group. In this way, the subspace W is given by $m + d_+^s - d_-^s - d$ linearly independent conditions, and its dimension is $d - (d_+^s - d_-^s)$. From this, $ind(B - I) = d_+^s - d_-^s$. $\qquad \square$

In the consideration of the spectral family of linear extensions of the form (15.1.4), depending on λ, all the introduced subspaces and corresponding dimensions also depend on λ. We find these quantities for the considered model operator (15.2.6) in the case when matrices $A_\pm$ satisfy the Perron condition.

Let eigenvalues $\lambda_\pm(k)$ of the matrix $A_\pm$ be numerated in the order of the increasing moduli, and let $L_\pm(k)$ be the corresponding one-dimensional subspaces. Recall that one only needs to consider those λ, for which $|\lambda| \neq |\lambda_\pm(k)|$.

We have

$$L_+^s(\lambda) = \bigoplus_{|\lambda_+(k)|>|\lambda|} L(k)_+, \quad L_+^u(\lambda) = \bigoplus_{|\lambda_+(k)|<|\lambda|} L(k)_+;$$

$$L_-^s(\lambda) = \bigoplus_{|\lambda_-(k)|>|\lambda|} L(k)_-, \quad L_-^u(\lambda) = \bigoplus_{|\lambda_-(k)|<|\lambda|} L(k)_-.$$

Denotes

$$d_\pm(\lambda) = \dim L_\pm^s(\lambda).$$

In order to find numbers

$$d(\lambda) = \dim[L_+^s(\lambda) \cap L_-^u(\lambda)]$$

one needs more delicate arguments, using a description of the dynamics of a linear map from the Theorem 10.2.1.

As it can be seen from the proof of the Theorem 15.2.5, the space $L_+^s(\lambda)$ is determined by the action of negative powers of the matrix A_+:

$$L_+^s(\lambda) = \{\xi : \lambda^n A_+^{-n}\xi \to 0\}.$$

Therefore, we consider the pyramid of the form (2.2.3), corresponding to the matrix A_+^{-1}. It is the pyramid consisting of subspaces

$$M_+(k) = \bigoplus_{j \geq k} L_+(j).$$

The subspace $L_-^u(\lambda)$, is determined by the action of positive powers of the matrix A_-:

$$L_-^u(\lambda) = \{\xi : \lambda^{-n} A_-^n \xi \to 0\}.$$

The pyramid, corresponding to the matrix A_-, consists of subspaces

$$M_-(k) = \bigoplus_{j \leq k} L_-(j).$$

Both of the constructed pyramids are complete. Let $k_\pm(x)$ are the exponents corresponding to these pyramids. Note that $k_\pm(x)$ is given as the index of the level of the pyramid $M_\pm(k)$, containing the vector x.

According to the Lyapunov Theorem, for these pyramids, there is a common normal basis $\{e(j)\}$, and we numerate the elements of this basis in such a way that $k_+(e_j) = j$, i.e. we assume that this basis is ordered with respect to the pyramid $M_+(k)$. In particular, $e(j) \in L_+^s(\lambda)$ if and only if

$$k_+(e_j) \le m - j$$

and the dimension of the subspace $L_+^s(\lambda)$ is the number of j for which this condition holds.

Let us now consider the values of the exponent k_- on this basis: let $s(j) = k_-(e_j)$. By the normality of the basis, all values $s(j)$ are different, and, consequently, the map $s : j \to s(j)$ is a permutation of elements of the set $\{1, 2, \ldots, m\}$, characterizing the relative position of two considered pyramids.

By the construction, we get that

$$\|A_-^n e(j)\| \sim |\lambda_-(s(j))|^n.$$

Therefore, $e(j) \in L_-^u(\lambda)$ if and only if

$$|\lambda_-(s(j))| < |\lambda|,$$

and the vectors $e(j)$, for which this condition holds, span the whole space $L_-^u(\lambda)$. In this way,

$$d(\lambda) = |\{j : k_+(e_j) \le m - j, |\lambda_-(s(j))| < |\lambda|\}|. \tag{15.2.10}$$

Let us formulate the obtained result.

Theorem 15.2.6 *Suppose that the coefficients $A_\pm$ of the model discrete weighted shift operator B satisfy the Perron condition. The operator $B - \lambda I$ is Fredholm if and only if*

$$|\lambda| \ne |\lambda_\pm(k)|.$$

Under this condition

$$\dim\ker(B - \lambda I) = d(\lambda), \quad \dim\ker(B^* - \bar\lambda I) = d(\lambda) - d_+(\lambda) + d_-(\lambda),$$

where $d(\lambda)$ is given by the formula (15.2.10). In particular, the operator $B - \lambda I$ is invertible if and only if $d_+(\lambda) = d_-(\lambda)$ and $d(\lambda) = 0$.

The arguments from the proof of the Theorem 15.2.6, also give the solution of the problem on the reduction of the operator to the normal form. Recall that, in this problem, one has to find the simplest form to reduce a weighted shift operator using transformations of the form

$$B \rightarrow S(x)BS(x)^{-1},$$

where $S(x)$ is a continuous invertible matrix-function. In the considered case, continuous matrix-function on the space $X = \mathbb{Z} \bigcup \{\pm\infty\}$ is a sequence of matrices $S(i)$, having limits at infinity $S(\pm\infty)$.

Theorem 15.2.7 *Suppose that matrices $A_{\pm}$, which are coefficients of the model discrete weighted shift operator B, satisfy the Perron condition. There is a continuous invertible matrix-function S on X such that in the weighted shift opertator $\widetilde{B} = SBS^{-1} = \widetilde{A}T$, the coefficient $\widetilde{A}$ is a diagonal matrix*

$$\widetilde{A} = \begin{pmatrix} a_1(i) & 0 & \cdots & 0 & 0 \\ 0 & a_2(i) & \cdots & 0 & 0 \\ \cdots & \cdots & \cdots & \cdots & \cdots \\ 0 & 0 & \cdots & a_{m-1}(i) & 0 \\ 0 & 0 & \cdots & 0 & a_m(i) \end{pmatrix},$$

where

$$a_j(+\infty) = \lambda(j)_+, \quad a_j(-\infty) = \lambda(s(j))_-,$$

and $s : j \rightarrow s(j)$ is a permutation of elements of the set $\{1, 2, \ldots, m\}$, characterizing the relative position of two considered pyramids.

Proof Let $\{e(j)\}$ be a previously constructed common normal basis for corresponding pyramid. For each j, $j = 1, \ldots, m$, the sequence

$$\frac{1}{\lambda(j)_+^n} A_+^n e(j)$$

has a limit as $n \rightarrow +\infty$, and this limit is an eigenvector of A_+, correspronding to eigenvalue $\lambda(j)_+$. For each j, $j = 1, \ldots, m$, the sequence

$$\frac{1}{\lambda(j)_-^n} A_-^n e(j)$$

has a limit as $n \rightarrow -\infty$, and this limit is an eigenvector of A_-, corresponding to eigenvalue $\lambda(s(j))_-$.

The matrix $S(n)$ can be constructed as the transition matrix to the basis

$$\frac{1}{\lambda(j)_+^n} A_+^n e(j), \quad j = 1, \ldots, m,$$

for $n \geq 0$, and as the transition matrix to the basis

$$\frac{1}{\lambda(j)_-^n} A_-^n e(j), \quad j = 1, \ldots, m,$$

for $n < 0$.   □

A proposition, analogous to the Theorems 15.2.6 and 15.2.7 for more general discrete weighted shift operators in spaces $l_2(\mathbb{Z}, \mathbb{C}^m)$ have been obtained in [7], but in this paper, the existence of the permutation $s(j)$ have been established only; in the model example, this permutation may be constructed explicitly by using linear algebra operations.

References

1. Antonevich, A. B. (1988). *Linear functional equations. Operator approach* (English transl. Birkhauser, Basel, Boston, Berlin, 1996). Minsk: Universitetskoe (in Russian).
2. Antonevich, A., & Lebedev, A. (1994). *Functional differential equations: I. C*-theory*. Harlow: Longman Scientific & Technical.
3. Chicone, C., & Latushkin, Yu. (1999). *Evolution semigroups in dynamical systems and differential equations*. Providence, RI: AMS.
4. Antonevich, A., & Lebedev, A. (1998). Functional and functional-differential equations. AC^*−algebraic approach (English transl. in Amer. Math. Soc. Transl. Ser. 2. 1999, pp. 25–116). *Trudy Sankt-Peterburgskogo Matematicheskogo Obshchestva, 6*, pp. 34–140.
5. Latushkin, Y. D., & Stepin, A. M. (1991). Weighted translation operators and linear extensions of dynamical systems (English transl. in Russian Math. Surveys, V. 46, no. 2, pp. 93–165). *Uspekhi matematicheskikh nauk, 46*(2), 85–143.
6. Antonevich, A. B. (2005). Coherent local hyperbolicity of a linear extension and essential spectra of a weighted shift operator on a closed interval. *Functional Analysis and Its Applications, 39*(1), 9–20.
7. Antonevich, A. B. (1978). A factorization of difference operator. *Vesti Akad. Navuk BSSR, Ser. Fiz.-Mat. Navuk, 3,* 10–15.

Brief Comments on the References

1. Among the extensive references on the Theory of Dynamical Systems, let us point out to books [1–4], containing the necessary material. The Theorems, expanded in the Sect. 1.3, are contained, for example, in [5–7]. Theorem 1.3.3 has been communicated to the authors by P. P. Zabreiko.

2. The Propositions from Sect. 2.1 are contained in textbooks on Matrix Theory, for example, [8, 9].

3. The Theory of exponents is thoroughly presented in [10]. On the theory of non-archimedean norms there are many sources, here we point out to [11–14]. We remark that in the general normalization theory [14] one considers norms with values in a linearly ordered group, i.e. the generalization of norms. The authors have not been able to find publications dedicated to the study of graded-linear operators.

4. The Expansion of vector trajectories have actually been written down in [15], but only the principal terms of the expansion were analyzed. The complete exposition is obtained in [16, 17].

5. The problem concerning the behavior under the action of a unitary operator and under the action of the standard shift on a torus, have been studied from different points of vies, since this problem is related to many other problems. Among them, the Lagrange problem on the average motions, the problem on the joint approximation of a collection of numbers by rational numbers, the problem of small denominators, the problem on the uniform distribution of number's sequence etc. Among the sources on this problem, we mention the books [1, 2, 18–20].
   The Theorems 5.2.2 and 5.3.1 are obtained in [15].

6. The Vector bundle theory is presented, for example, in [21–23]. The Theorem 6.6.1 is published in [24].

7. The Grassmann manifolds appear in various subjects and their description is contained not only in books on geometry; see, for example, [9, 25, 26] . On metrics on the Grassmann manifolds, see [9].

8–9. For the exterior powers of spaces and operators, see [1, 9, 27].

10. The Theorem 10.2.1 is published in [28], the strengthened version in [29].

© Springer International Publishing Switzerland 2014
Ć.B. Dolićanin and A.B. Antonevich, *Dynamical Systems Generated by Linear Maps*,
DOI: 10.1007/978-3-319-08228-8

11. The Theorem 11.1.1 is contained in [17, 30].
12. The Theorem 12.5.1 is published in [31].
13. The Theorem 13.1.1 is published in [32], detailed exposition is contained in [33, 34]. The formula for the spectral radius was obtained independently, and almost at the same time, in [35]. For some of concrete classes of operators, the claim of the Theorem 13.1.1 was obtained earlier, together with the description of ergodic measures, in [36, 37].
14. The Theory of C^*-algebras is presented, for example in [38–40]. The approach towards investigation of operator algebras, induced by dynamical systems, among them the algebra of differential-functional operators, based on the theory of C^*-algebras, expanded in [33, 34, 41].
15. The results from 14.3.2 are obtained in [42].

References

1. Katok, A., & Hasselblatt, B. (1998). *Introduction to the modern theory of dynamical systems*. Cambridge: Cambridge University Press.
2. Kornfeld, I. P., Sinai, Ya. G., & Fomin, S. V. (1980). *Ergodic theory*. Moscow: Nauka.
3. Halmos, P. (1956). *Ergodic theory*. New York: Chelsea.
4. Bronshtein, I. U. (1984). *Nonautonomous dynamical systems*. Kishinev: Shtiinsa.
5. Lankaser, P. (1969). *Theory of matrices*. New York: Academic Press.
6. Bellman, R. (1960). *Introduction to matrix analysis*. New York: McGraw-Hill Book Company.
7. Voevodin, V. V., & Kuznetsov, Yu A. (1984). *Matrix and calculations*. Moscow: Nauka.
8. Gantmakher, F. R. (1988). *Theory of matrices*. Moscow: Nauka.
9. Glazman, I. M., & Lubich, Yu I. (1969). *Finite-dimensional analysis problems*. Moscow: Nauka.
10. Bylov, B. F., Vinograd, R. E., Grobman, D. M., & Nemyckii, V. V. (1966). *Lyapunov exponents*. Moscow: Nauka.
11. Khrennikov A. Yu. (2003). *Non-Archimedean analysis and its applications*. Moscow: FizMatLit.
12. Mahler, K. (1980). *Introduction to p-adic numbers and their functions*. Cambridge: Cambridge University Press.
13. Radyna, A., Radyna, Y., Radyna, Y. (2010). *Basis of non-archimedean analysis*. Minsk: Belarussian State University Publications.
14. Zariski, O., & Samuel, P. (1960). *Commutative algebra* (Vol. II). Princeton: D. Van Nostrand com. INC.
15. Antonevich, A., & Buraczewski, A. (1996). Dynamics of linear mapping and invariant measures on sphere. *Demonstratio Math.*, 29(4), 817–824.
16. Reshic, S. (2006). *Asymptotic expansion of vector's trajectory under action of a linear mapping*: Trudy Conference Mathematical Models and Boundary Problems, Samara, pp. 34–37.
17. Antonevich, A., Dolichanin, Ch., & Reshic, S. (2008). On convergence of a trajectory of a vector subspace: Case of single eigenvalue. *Doklady NAN Belarusi*, 52(1), 27–32.
18. Weyl, H. (1984). *Mathematics. Theoretical physics*. Moscow: Nauka.
19. Arnold, V. I. (1978). Supplementary chapters of the theory of ordinary differential equations. Moscow: Nauka. (English Trans. *Geometric methods in the theory of ordinary differential equations* (2nd Ed.). New York: Springer (1989)).
20. Shidlovskii, A. B. (1982). *Diophantine approximations and transcendental numbers*. Moscow: MGU Publcations.
21. Atiyah, M. F. (1967). *K-theory*. New York: W.A. Benjamin.

22. Husemoller, D. (1966). *Fibre bundles*. New York: McGraaw-Hill Book Company.
23. Mishchenko, A. S. (1984). *Vectors bundles and its applications*. Moscow: Nauka.
24. Antonevich, A., & Dolichanin, Ch. (2010). *Foliation of the set of invariant subspaces: Proceedings of 5th International Conference on Analytical Methods of Analysis and Differential Equations* (Vol. 2, pp. 12–16) September 14–19, 2009, Minsk.
25. Dubrovin, B. A., Novikov, S. P., & Fomenko, A. T. (1979). *Geometry, modern, methods and applications*. Moscow: Nauka. (English transl. *Parts I, II, III*. New York:Springer, 1984, 1985, 1990).
26. Rokhlin, V. A., & Fuks, D. B. (1977). *Introduction to topology. Geometric chapters*. Moscow: Nauka.
27. Lang, S. (1965). *Algebra*. Reading: Addison-Wesley Publications.
28. Antonevich, A., Dolichanin, Ch., & Nicolich, G. (2001). Dynamics of a linear mapping on the Grassman manifold. Perron's case. *Trudy IM NAN Belarusi, 9*, 20–23.
29. Antonevich, A., Dolichanin, Ch., & Nicolich, G. (2001). Dynamics of a linear mapping on the Grassman Manifold. Case of different eigenvalue. *Spectral and Evolution Problems, Simferopol, 11*, 70–72.
30. Reshic, S. (2007). *Trajectory of a vector subspace: case when absolute value of eigenvalues are equal: Trudy Conference*. Minsk: Eruginskie chteniia.
31. Reshich, S., Antonevich, A. B., & Dolichanin, Ch. (2007). Convergence of a trajectory of a vector subspace under the action of a linear map: General case. *Novi Sad Journal of Mathematics, 37*(2), 149–159.
32. Lebedev, A. (1979). Invertibility of elements in C^*-algebras generated by dynamical systems. *Uspekhi Mat. Nauk, 34*(4), 199–200.
33. Antonevich, A. B. (1988). *Linear functional equations. Operator approach*. Minsk: Universitetskoe. (English trans. in Russian, Birkhauser, Basel, Boston, Berlin, 1996).
34. Antonevich, A., & Lebedev, A. (1994). *Functional differential equations: I.C^*-theory*. Harlow: Longman Scientific & Technical.
35. Kitover, A. (1979). The spectrum of automorphisms with weight, and Kamowitz-Scheinberg theorem. *Funktsion. Anal. i Prilozhen. 13*(1), 70–71 (English Trans. in Russian in *Functional Analysis and Its Applications*. 13 (1979)).
36. Antonevich, A. B. (1975). On a class of pseudodifferential opearators with deviating argument on the torus. *Differentsial'nye Uravnenija, 11*(9), 1550–1557.
37. Antonevich, A. B. (1979). Operators with a shift generated by the action of a compact Lie group. *Sibirsk. Mat. Zh., 20*(3), 467–478.
38. Murphy, G. J. (1990). *C^*-Algebras and operator theory*. San Diego: Academic Press Inc.
39. Dixmier, J. (1969). *Les C^*-algèbres et leurs Représentations*. Paris: Gauthier -Villars.
40. Gelfand, I. M., Raikov, D. A., & Shilov, G. E. (1960). *Commutative normed rings*. Moscow: FizMatGiz.
41. Antonevich, A., Lebedev, A. (1998). Functional and functional-differential equations. AC^*-algebraic approach. Trudy St.Peterby. *Mat. Obshch., 6*, 34–140. (English Trans. in *American Mathematical Society Translations: Series 2*, 25–116, (1999)).
42. Pietruczuk, B. (2010). *Jednostronna odwracalność operatora B − λI, generowanego przez odwzorowanie liniowe*. Uniwersytet w Bialymstoku: Praca magisterska.

Bibliography

43. Antonevich, A. B. (2005). Coherent local hyperbolicity of a linear extension and essential spectra of a weighted shift operator on a closed interval. *Functional Analysis and its Applications, 39*(1), 9–20.
44. Danes, J. (1987). On the local spectral radius. *Cas. Pest. Mat., 112*, 177–187.
45. Marcus, M., & Minc, H. (2004). *A survey of matrix theory and matrix inequalities*. Editorial Moscow: URSS (Russian translation).

Printed in the United States
By Bookmasters